Sustainability and Design Ethics

Sustainability and Design Ethics

Second Edition

Jean Russ

CRC Press is an imprint of the
Taylor & Francis Group, an **informa** business

CRC Press
Taylor & Francis Group
6000 Broken Sound Parkway NW, Suite 300
Boca Raton, FL 33487-2742

© 2019 by Taylor & Francis Group, LLC
CRC Press is an imprint of Taylor & Francis Group, an Informa business

No claim to original U.S. Government works

Printed on acid-free paper

International Standard Book Number-13: 978-1-1385-8368-9 (Paperback)
International Standard Book Number-13: 978-1-1383-9041-6 (Hardback)

This book contains information obtained from authentic and highly regarded sources. Reasonable efforts have been made to publish reliable data and information, but the author and publisher cannot assume responsibility for the validity of all materials or the consequences of their use. The authors and publishers have attempted to trace the copyright holders of all material reproduced in this publication and apologize to copyright holders if permission to publish in this form has not been obtained. If any copyright material has not been acknowledged please write and let us know so we may rectify in any future reprint.

Except as permitted under U.S. Copyright Law, no part of this book may be reprinted, reproduced, transmitted, or utilized in any form by any electronic, mechanical, or other means, now known or hereafter invented, including photocopying, microfilming, and recording, or in any information storage or retrieval system, without written permission from the publishers.

For permission to photocopy or use material electronically from this work, please access www.copyright.com (http://www.copyright.com/) or contact the Copyright Clearance Center, Inc. (CCC), 222 Rosewood Drive, Danvers, MA 01923, 978-750-8400. CCC is a not-for-profit organization that provides licenses and registration for a variety of users. For organizations that have been granted a photocopy license by the CCC, a separate system of payment has been arranged.

Trademark Notice: Product or corporate names may be trademarks or registered trademarks, and are used only for identification and explanation without intent to infringe.

Visit the Taylor & Francis Web site at
http://www.taylorandfrancis.com

and the CRC Press Web site at
http://www.crcpress.com

For my good friend Thomas G. Mudra

and always to Karla

Contents

Preface ix
Preface to the Second Edition xiii
Acknowledgments xv
Acknowledgments for the Second Edition xvii
Author xix

1. Why Does Prometheus Suffer? 1
Sustainability and Design 2
   Unsustainable 2
   What Is Sustainability? 3
Why Sustainability Is Needed? 9
   Selling Sustainability 12
Endnotes 16

2. Professional Ethics 19
What Is a Profession? 19
   "Due Diligence" in Codes of Ethics 23
   Assessing Due Diligence 23
Evaluating Codes of Ethics 30
Sustainability and the Ethical Challenges for Designers 34
Bibliography 44
Endnotes 45

3. Is There an Ethical Obligation to Act Sustainably? 49
Theories of Ethics 49
   What Obligations Do We Have to Other Living Things? 54
   Utilitarian Views of Nature 55
   Speciesism 57
Who Owns the Environment? 60
   Nature as Property 63
   The Value of Property and Nature 64
   Reconciling Private Property and Sustainability 65
   Do No Harm 67
Role of Professional Standards 68
Endnotes 69

4. The Design Professional and Organizations 73
Balancing Obligation and Opportunity 73
   Can There Be Deeds Without Doers? 76
   Ethical Agency 77
The Descent of Corporate Obligation 79
   Business and Sustainability 83
   Growth in Sustainable Business Practices 85
   Why Should Corporations Be Viewed as Having Moral Responsibility? 88
   The Duty of the Corporation to Professional Employees 89
Justification for a Whistle-blower 96
Endnotes 100

5. The Choice for Sustainability 107
The Design Professional as Leader 108
Standards Cannot Produce a Sustainable World 110
Sustainability and Obligation 116
Science and Design 118
Sustainability and Design Ethics 122
Endnotes 125

6. The Precautionary Principle and Design 127
Moral Underpinnings of the Precautionary Principle 129
Precautionary Principle and Design 135
   What Do We Owe the Future? 137
   Duty to Prevent Harm 138
Precaution and Design 140
Endnotes 141

7. Flourishing 145
What Ought We Do? 147
Designer as Teacher 149
Design Values 151
Designer as Student 157
Endnotes 159

Appendix A 161

Appendix B 171

Index 175

Preface

If you have picked up this book based on the title, you are probably interested in or involved with design in some capacity. Most people of my acquaintance who are so involved have thought about what sustainability means in terms of their work and whether and how to embrace it. There is some reticence to being on the "bleeding edge" of new approaches and thinking when you are in the business of design. So many designers and design practices wait and plan to arrive with sustainable design services when there is market demand. That time may be upon us. The growth in interest and demand for energy efficiency in homes and commercial buildings has grown dramatically in only the last few years, and the number of "green buildings" has followed suit. Still, these buildings represent a very small portion of the actual buildings completed each year. The demand for "green" products has also increased, so that, in many cases, it is difficult to determine what green actually means. These may be signals of the shift in marketplace, or they may be moments of popularity that will fade with the next great idea. I believe they are the former.

It is not that as a society we will have gained some sort of transcendent wisdom when it comes to the environment, although there is clearly some awakening in terms of our environmental consciousness. Rather, we will embrace "green" because it makes economic sense to do so and because the market will respond to the economic demand. Faced with increases in climate uncertainty, rising energy costs, and the pressures on food, water, and services from ever-greater world and national populations, we will adapt. I believe many of those adaptations will be centered around what we call sustainable design, because the sustainable argument simply makes more sense when all of the costs and benefits are tallied up.

When work on this project started, I focused on design processes with which I was familiar. I am a landscape architect, and so, I am more familiar with site design and landscape restoration processes. As I worked through writing the book, however, I realized that the reasoning applies to any design process and any design profession. At the heart of design is a core or foundational set of desires and values. This book is concerned with a particular aspect of professional design—the ethical obligations that sustainability will create for designers. In my view, it is part of a conversation between design professionals that constitutes an important part of the evolution of professional standards of care. It is my intention that this book contributes something to the long line of books about design ethics but particularly to that part of it written by designers themselves, such as Raymond Spier, Edwin Layton Jr., Samuel C. Florman, Tom Spector, Victor Papanek, Thomas Fisher, William McDonough, and Michael Braungart, and others. The nature of professional ethics is necessarily dynamic, reflecting changes in science, society, materials, and methods, as well as changes in designers themselves. It is the premise of this book that we are in a period of significant challenges and changes that are compelling changes in how we think about design and what our expectations of outcomes will be. The ethical component of those changes and challenges has elements of being both cause and effect.

I believe one must approach the subject of ethics carefully. There is a sense that ethical questions and issues occur in two spaces: in a personal space and on a professional stage. The latter is subject to discussion and criticism, but the former tends to be off limits. This book is concerned with the ethical challenges for the professional acting in the marketplace, on the pubic stage. There is, however, a sense that the professional is not divorced from his personal ethics and values. So, in that sense, it is concerned with the values we bring to our work and how these are expressed in our professional conduct. Professional and personal ethics are rooted in moral reasoning—whether or not we engage in them, they are subject to rational analysis. When we analyze our ethics, we look at the moral reasoning on which they are based and the rationality of that reasoning. Since this is true, it is possible to have a rational discussion of how a change in our understanding of the world will result in a change in our ethical obligations and behavior.

The need for sustainable outcomes from design is apparent in the myriad of examples we see every day: from food that is unhealthy to eat, to unsafe children's toys, to containers that leach toxins into what we eat and drink, and from products and systems that waste energy and pollute the environment, to community design that leads to increases in poor health, and so on. If the process of design is to make manifest the ideas of human creativity and desire, what do these outcomes say about designers? What values underlie these designs? The growing demand for sustainable outcomes will require us to rethink the outcomes expected from our work, and that, in turn, requires us to rethink the values and ethics that guide our designs. This book is an attempt to conduct this value analysis and to make a reasoned argument for new design values and ethics.

Chapter 1 makes an argument for sustainability and links sustainability to the design professions, both as an obligation and as an opportunity. The idea of sustainability is explored in terms of just what it means in general but more particularly what it means to a designer. Some of the various underlying ideas of how to define sustainability are discussed. The chapter includes some discussion on what unsustainability is, by considering the results and impacts of current approaches to design in various fields. It includes the introduction of several ideas and concerns, which are the subject of subsequent chapters.

Chapter 2 is a fairly broad exposition of codes of professional ethics. Codes of ethics of selected professional groups are reviewed in the context of how they have evolved to include issues of environmental concern and whether they might meet some tests of due diligence. The chapter explores the idea of a profession as distinct from other forms of work and career. The nature and context of professional obligations are considered. The review of the various codes of ethics considered reveals some important information. There is some consistency across the organizations reviewed, but it is the differences where interesting facts emerge. Some professions have moved down the path toward including sustainability within the scope of their code of ethics, while others are silent on the issue. This suggests that progress is being made but that we are still in the early stages of adopting a sustainability ethos. This discussion leads to a more defined discussion in Chapter 3 as to whether sustainable design is an ethical obligation for professionals. Can sustainability be considered an issue of public health, safety, and welfare? If it can be, what obligations does that create? How are elements of nature and the environment to be evaluated in the design process? Would an obligation to design for sustainable outcomes create an obligation to nonhuman life, to other species? Chapter 3 attempts to provide rational and moral reasoning for including sustainable outcomes within the ethical considerations of design professionals.

Recognizing that most design professionals conduct their work within organizations as employees, Chapter 4 considers how this might influence the individual designer's ethics and obligations in general. The chapter is concerned with the professional as an employee and as a supervisor of other professionals. Since within organizations and on large projects an individual designer may only be involved with a small portion of the total project, how does one make ethical decisions? Where does the obligation exist, and where are there limits on ethical duty? A fair portion of the chapter is concerned with the ethical obligation of the professional corporation or the corporation providing design services. Can a corporation have ethical obligations? What obligations does a supervisor or corporation have to the professional employee? In the end, if faced with intractable circumstances, what considerations should the professional make before making a decision to "blow the whistle" on an employer.

Chapter 5 is concerned with focusing on the arguments made in preceding chapters on the choice the individual professional and organization makes to pursue sustainable outcomes. It is concerned with how this might work in the real world of multiple stakeholders, often with competing interests, of regulations and design standards that are often decidedly unsustainable. As the leader of a design process, the design accepts particular obligations, but as an actor in the real world, where clients can choose from many professionals offering services, where there is reluctance to change by those charged with regulating activities, and where the design synthesis for a sustainable outcome may represent significant departure from past practices, balancing such obligations can be a challenge.

Since there will always be some degree of uncertainty in even the simplest design, how can a design professional meet the test of designing for a sustainable outcome? The manufactured and built environment is rich with examples of unintended consequences. It is certain that there will be unintended consequences in the future that will reduce or mitigate the designer's intent. How the individual designer can be certain that she conducted her work with the appropriate degree of diligence, the precautionary principle, and its application for designers are all explored in Chapter 6.

The final chapter of the book attempts to summarize the preceding chapters and suggest ways in which the various design professions might meet the ethical challenges to provide for sustainable outcomes in their work. In the end, all design is hopeful, and all designers are, at heart, optimists. In periods of change lie the greatest opportunities for individuals to distinguish themselves and for professions to demonstrate why they deserve the trust of the public. A key component of design is the underlying philosophy or values of the designer. Chapter 7 introduces the idea of what sorts of design values might facilitate sustainable design and result in desirable sustainable outcomes.

The purpose of this book is to contribute something to the deliberations that take place among designers and within organizations, as this period of adaptation proceeds. I am mindful that no one opinion on these matters can provide all of the answers. In one sense, the book is meant to raise as many questions as answers. It will hopefully provoke thoughtful consideration and discussion as we find our way into the new design paradigms that will lead to a sustainable future and successful design practices.

Preface to the Second Edition

This second edition is an update of many of the statements and claims made in the first edition. Some aspects of sustainability as a practice changed a good deal in the time between the first and second editions. As a concept, it remains as challenging and as important as ever, perhaps even more so. Items referenced in the text such as codes of ethics are necessarily dynamic, and so, references to them require periodic updates. The worlds of commerce and politics are also dynamic, and, to put it simply, things change. In the nearly 10 years that have passed since I worked on the first edition, major changes have occurred in the awareness and commitment of businesses and governments. The number of organizations with real, measurable commitments to sustainability remains small, but there has been real and meaningful growth in the number and influence of those businesses. Many national and local governments have set out to become sustainable as well. Unfortunately, there are still organizations for whom sustainability is a marketing effort and little more, and the leadership of the federal government of the United States has a decidedly backward approach to the idea of sustainability.

Besides updating some of the references and so on, this edition expands the discussion of the role of business in sustainability and its relationship to designers. The consideration of peak organizations has also been expanded to have a more international scope, including a look at peak organization in Asia, the European Union, and other parts of the world.

Acknowledgments

It is important to first thank and acknowledge Dr. Albert Anderson of Agora Press and the author of the paper "Why Prometheus Suffers: Technology and the Ecological Crises." Dr. Anderson's paper was among the things that helped the ideas that became this book take shape in my mind. His thoughtful emails in the first days of preparing the manuscript were helpful and encouraging. This book was prepared with the generous assistance of Robert B. Ludgate Sr., PE, PLS, Keith Seddon, PhD, Niall Kirkwood, and Zolna Russell, RLA. I very much appreciated the time and assistance provided by Mary Hanson from the American Society of Landscape Architects, Tera Hoke from the American Society of Civil Engineers, and Arthur Schwartz from the National Society of Professional Engineers.

Every author owes a debt of gratitude to those people who listen to the ideas and arguments as they are born half-baked and incomplete through the process of preparing a manuscript. Keith Seddon, PhD, provided valuable instruction, feedback, and criticism in my first serious reading in the field of ethics. I am grateful to Robert Ludgate Sr., PE, PLS, for his valuable mentorship years ago and his comments on the first draft material. I am also thankful to my colleagues at the College of Southern Maryland, Professor William Montgomery, Professor Richard Bilsker, and Professor Lee Vines, for their feedback and input. Without the opportunity provided by Taylor & Francis Group, this project would not have gone forward; my thanks to Irma Shagla for her confidence in the project and Jennifer Ahringer for her able direction and oversight in the production of this book. My friend Tom Mudra deserves particular appreciation for his challenging questions and advice, willingness to listen, and no-nonsense critiques. Finally, as always, I rely on my friend and insightful critic Karla Russ for her advice, companionship, and love.

This book was the product of many conversations, readings, and lessons over the years as I sought my own sense of the questions of professional responsibility and environmental integrity. In the end, it is a distillation of the work and ideas of many people, as processed in my notes and thinking. As such, while there is much credit to be given, the responsibility for any errors in the facts or arguments as presented is my own.

Acknowledgments for the Second Edition

This second edition builds directly on the work and the contributions of others that went into the first edition, and so, I must acknowledge their contributions once again. In this second edition, I must also acknowledge the assistance from Dr. Patrick Allen, Richard Bilsker, Robert Ludgate Sr., PE, George Gibson, and Xu Chunmei. Irma Shagla provided the opportunity and encouragement along the way, and without her and the staff at Taylor & Francis Group, this book would not have been possible. Thank you to James Gavacs and Stephen Russ for their work on the cover image for this edition.

And as always, I appreciate the love and support of Karla, my family, and my friend Tom Mudra.

Jean Russ

Author

Jean Russ is a graduate of the State University of New York, having completed the bachelor of science degree in environmental studies. She completed a master's degree from Kutztown University, Pennsylvania, after which she studied ethics in a doctoral program for some time. She is a registered landscape architect and the author of five books, including *Brownfield Redevelopment, The Site Planning and Design Handbook,* and the first edition of this book. As a design professional, Ms. Russ was a principal in several firms and acted as project manager in numerous development and brownfield projects. Ms. Russ currently serves as the chair of the Science and Engineering Division at the College of Southern Maryland.

1

Why Does Prometheus Suffer?

Art is simply a right method of doing things. The test of the artist does not lie in the will with which he goes to work, but in the excellence of the work he produces.

Thomas Aquinas

In Greek mythology, Prometheus plays the important role of bringing the technical arts to humans in the form of fire. Dr. Albert Anderson revisits this story from Plato's Protagoras in a paper entitled "Why Prometheus Suffers: Technology and the Ecological Crises"[1] and finds important parallels from the story of Prometheus and our contemporary concerns with ecological crises and sustainability. In the myth, the Titans, Epimetheus, and Prometheus (Epimetheus means "afterthought" and Prometheus means "forethought") are given the task of distributing the means of survival to each living creature. Epimetheus asks Prometheus that he be allowed to do it, and with Prometheus' agreement, he proceeds. Under Epimetheus, some creatures are given speed, some strength or cunning, and others flight, etc. When Prometheus returns to see how it is going, he discovers that Epimetheus has finished his task but there is nothing left to give to humans. He observes humans' discomfort: hungry, cold, and defenseless, and he pities them. So, Prometheus steals fire from Olympus and with it the technical arts for humans. With these gifts, so Plato tell us, we acquired the ability to know things, to build, to engineer, and to use letters that allow us to "hold all things in memory." However, these gifts are not distributed equally among all people and therefore some are good with trade work, others are good with smithing, others are skilled at war, and so on.

Zeus observes humans with their new capabilities and finds them to still be wanting; they are unorganized and without virtue. Hermes is directed by Zeus to go to Earth and to give people the civic arts "to bring respect and right among men, to the end that there should be regulation of cities and friendly ties to draw them together." Hermes inquires if these civic arts should be distributed as were the arts stolen by Prometheus (one given the medical arts, another carpentry, a third farming, and so on). Zeus says, no, "Let them all have their share; for cities cannot be formed if only a few have a share of these as of other arts."

Zeus punishes Prometheus by chaining him to a rock at the top of a mountain. Each day, Prometheus is visited by Zeus in the form of an eagle, who proceeds to tear at and devour his liver. Each night, he is restored, only to be tormented again the following day. Why does Zeus punish Prometheus? Prometheus suffers because of the problem created by giving men fire and technology but not the civic virtue to complement them. Zeus understood that if there is to be balance and good in society, then the civic arts must accompany the technical arts. Prometheus is punished because, as having the gift of forethought, he should have known that by stealing only the technical gifts, humans would become "A menace to themselves, to other creatures and to the earth itself."

Professor Anderson asserts in his paper and in a new translation of the Protagoras dialogue[2] that our crisis is not one of technology but rather one of philosophy. Prometheus suffers because he had the capability to know better but did not act on it. We have tended to see and respond to ecological crises and environmental imbalance with an Epimethean philosophy, that is, as an afterthought rather than with forethought. What is required, perhaps, is a Promethean Philosophy of looking forward.

Sustainability and Design

Our society is in the process of recognizing that our relationship with nature and the environment must change. For some, it seems as if this recognition is terribly slow in coming, if indeed it is even coming along quick enough. For others, the pace and concerns driving it seem to be rushing forward, even threatening to the familiar old ways of our experience. Whether the rate of change is too slow or fast is probably best judged from some future perspective, but that significant change is underway, necessary, and widely acknowledged. Truly, some are in a position to better see the need and observe the pace of change. Many indigenous people and those with the training and interest to measure the losses to the diversity and quality of the environment have spoken out for decades, perhaps longer. As the data has accumulated and the observations have become more obvious and easier to detect, more of us have acknowledged that, indeed, our decisions and actions have significant and detrimental implications on the environment. The scope and scale of the problems are so large and daunting, many are uncertain about what they might do in the face of this knowledge. Design professionals of all stripes are among those best suited by education and profession to provide answers to the questions and solutions for the problems. However, defining just what sustainability is can be problematic. In some ways, it may be like quality, hard to define, but you tend to know it when you see it. Still, an alternative to business as usual is called for.

Unsustainable

"Sustainable America—A New Consensus for Prosperity, Opportunity and a Healthy Environment for the Future" was published by the President's Council on sustainable development in 1996.[3] The report identified 10 goals of sustainable development, but the first three could be viewed as the most important: (1) health and the environment, (2) economic prosperity, and (3) equity. In this context, equity refers to social equity (equal opportunity) and intergenerational equity (equity for future generations). It is widely recognized that to meet these goals, we must change the way we behave. In a very real sense, economic prosperity has been pursued with less than commensurate attention to social equity and environmental sustainability. Indeed, at times, there is an outright antagonism between those advocating "progress" and those voices speaking for social equity and environmental sustainability. Adopting paradigms of sustainability will require us to reconcile our economic interests with our environmental interests and social inequities.

Unfortunately, since the start of the President's Council on Sustainable Development (PCSD) in 1996, results have been uneven at best. President Obama tried to lift the council from its fairly moribund state in 2012 by committing the US federal government to various goals and initiatives, which resulted in some successes, but the actual results of that effort were spotty. Among the most important and visible steps toward sustainability has been the growth in wind and solar power generation, which resulted from federal and state support as well as from changes in the cost of solar equipment.

From the beginning, the Trump administration has worked against sustainability on most fronts. Appointees to the Environmental Protection Agency and Departments of Energy and Interior proudly and aggressively pursue an anti-environment and anti-sustainability mission requiring business and public interest groups to go to court again and again to protect gains made in the recent past. If there is good news in current events, it is in the pale hope for better days. A more substantive hope might be seen—after the 2016 election, the members of the Corporate Eco Forum convened a closed roundtable discussion and concluded that the current climate in Washington, DC, will not affect their moves toward sustainability.[4] A total of 60% of businesses responding to a survey in 2016 indicated that the changes in administration would not impact their sustainability goals, and another third admitted that it would slow them down, but progress would continue.[5]

What Is Sustainability?

There is a conversation within the design and planning professions that has been underway in one form or another for more than two decades. The conversation revolves around several concerns. First, what is sustainability and how is to be "sold" to clients and stakeholders? Perhaps the most common definition for sustainability was crafted in 1987 at the World Commission on Environment and Development, often referred to as the Brundtland Commission, "Development that meets the needs of the present without compromising the ability of future generations to meet their own needs." This definition captures the concept in broad strokes that are difficult to translate into action. In the final analysis, sustainability is complex, perhaps irreducibly so, but does it follow that sustainable design is also complex? The sustainable design is, in one sense, not an attempt to recreate nature from scratch but to mitigate the impacts of our actions on the existing system.

To some extent, the term sustainability has become almost meaningless. We may share a general sense of what we think we mean by sustainable, but there may be considerable difference of opinion when the details are to be thrashed about, and, of course, design is about details. Design is a process of bringing the imagined into reality. It is art and science, it is technical and creative, it is practical, and it is beautiful. It does not thrive or produce its best result when constrained by preset conditions. Ultimately, if the products of design are to be valuable, they must serve the interests of society. In this sense of "value," the products of designers are a reflection of our more fundamental values. Professor Anderson was correct in his paper when he posited that the ecological crisis that he was concerned about was not a failure of technology but rather a failure of philosophy. In fact, we know how to design "sustainably" for the most part, but we do not do so for a variety of reasons. Even with those reasons, we now understand that if the product of our work treats nature as an afterthought, we do harm ourselves, others, and nature largely because we value other things more. If we intend to design the elements of a sustainable world, then we must first have the values that direct such work. The design will reflect our values. Perhaps, a Promethean ethic will be manifested after all; a sustainable society will have the underlying values that produce sustainability in its products and the built environment.

As the various definitions suggest that to become a sustainable society, we must balance the interests of the present with the interests of the future in such a way as to realize opportunity within the limitations of the environment. Environmental concern in the modern economy is commonly considered a luxury, less necessary, or even desirable when more fundamental (economic) outcomes are not being met. In this view, the environment and the economy are often viewed as adversarial by nature, environmental concerns are often portrayed in terms of the negative impact on economic concerns, and sustainability requires us to find a meaningful resolution between them. In point of fact, sustainability requires us to maintain a quality environment and a robust economy in other than the zero-sum approach of the past. A weak economy or a lack of opportunity invites environmental degradation, if only, because individuals will act in their own interests. When those interests are threatened or impacted, concern with the environment is reduced.

Abraham Maslow wrote that we are motivated to fulfill needs, starting with the most fundamental physiological requirements for air, water, and food and proceeding through needs for safety, love, and ultimately self-esteem.[6] We tend to focus on the more fundamental needs and move on to higher needs only when the more basic desires are satisfied. In the end, we must understand that sustainability requires a healthy growing economy and a robust economy can exist only within a healthy and diverse nature.

Modern developed societies are primarily economic constructs; in practice, we are *homo economicus,* perhaps more than we are *homo sapiens.*

The environment, or our current valuation of it, can be described in broad economic terms as capital and resources. If capital is wealth that we might invest to generate more wealth, then nature provides capital in the form of stocks of materials and services that can be used to generate wealth or from which we might draw utility. If managed sustainably, the supply of wood from forests, or fish from the sea, can provide a stream of products indefinitely. The forest also contributes oxygen to the atmosphere, pumps water into the air, and provides other important environmental "services." Likewise, a functioning wetland provides a range of services, from water treatment to flood buffering. Healthy natural landscapes invite economic activity in terms of tourism and outdoor recreation in addition to the array of natural services they provide. Like economic wealth, the idea is to use the natural capital to generate greater wealth but never to spend the capital itself.

Wealth is generated by investing and managing capital, but it is also generated by consuming resources. Resources are consumed to make new wealth. Oil used as fuel is consumed; once used, it is no longer useful as a resource. When we harvest forests or fish at a rate that is unsustainable, we are treating our natural capital as a resource. Eating the seed corn as it were reduces the natural capital available, which by definition is an unsustainable act. Natural resources are generally only considered valuable in terms of how they might be consumed. There is often little incentive to manage resources in such a way as to ensure continued, albeit smaller levels of harvest or exploitation. This is especially true in the cases of unowned or public resources such as fisheries and forests. In competitive marketplaces, the individual often has little incentive to conserve or limit consumption to sustainable levels in the interest of the future or the community at large. Instead, in the absence of a future benefit that outweighs current satisfaction, the incentive is to take as much as possible. Finding a means to incentivize conservation and resource management is necessary to find a willing partner in the fisherman and logger. Various attempts to provide this incentive have met with some success around the world. For example, dividing a fishery into shares owned by commercial fishermen changes their relationship to the resource in a fundamental way. Earlier, the resource was not owned by anyone, and so, it was in the short-term interest of every fisherman to maximize his take. If the fishery is owned in shares and the fisherman is entitled to a predetermined share of the take, it is in his interest to have the fishery grow, so his share is larger. Early tests of this approach have shown promise.[7]

In the end though, sustainability expands our view of nature as capital to be invested and limits the use of nature as a resource whenever possible. This view of nature as capital or resource is at the heart of the rational argument for sustainable design. It is clear that resources are generally finite in nature. Even renewable resources can be considered as finite if the rate of consumption exceeds the sustainable yield, as it often does. Consumption, even at a nominal rate, by a growing population of more than 6.5 billion people, can quickly overwhelm the rate at which a given resource can replenish itself. Thomas Malthus anticipated in his essay "An Essay on the Principle of Population" that a growing population would increase at a rate faster than food supply.[8] He observed that while populations might increase exponentially, the growth of agricultural production was arithmetic or linear. Within some time, the number of people would be greater than the ability of a society to feed them. He also observed that as populations grew, the availability of surplus labor would drive wages down. More recently, Jared Diamond has written how societies sometimes consume their resources to such a degree that it leads to or contributes to a failure of the society as a whole. In part, sustainability is about resource management and protection and moving some natural resources into the natural capital account.

The issue is often described as one of consumption; if we simply use less stuff, the world will be a better place. That might be true, but it might just as well be true it is not simply the amount of consumption but the resources and processes necessary to facilitate what we consume. Consumption is natural, and everything consumes in the natural order of existence. Consumption in and of itself cannot be bad, but there is no free lunch either. If we allow ourselves to consume our way to failure, then we shall fail. There is ultimately a finite supply of material things. Certainly, technology will enable to adopt new methods to use new materials to avoid consequences of flagging supplies of a resource. The Green Revolution utilized new methods and agricultural inputs and cheap energy to produce more food. It now widely agreed that those techniques have some significant costs and become increasingly more expensive to maintain. Perhaps, genetically modified food is the next Green Revolution. But who can know at this point? Consumption will continue, but sustainability requires that we recognize all of the costs and mitigate the effects of unsustainable consumption.

Sustainability has also been described as occurring on a continuum between weak-side and strong-side sustainability. Weak-side sustainability is also sometimes called the "constant capital rule." In this view, an act is considered sustainable as long as the amount of capital in a system remains the same; in other words, we are able to shift capital between categories. Capital can be found in various forms, ranging from human capital (creativity and problem solving) to economic capital to natural capital. The weak side allows capital to be shifted from natural capital to manufacturing capital or equity in a real-estate development project. So, we can offset losses in natural capital with concomitant increases in other capital accounts. A loss in air quality might be sustainable if the economic value of the infrastructure that caused it were to create an offsetting lasting positive value. In such an instance, the total capital in the system remains the same for the future. This would not be the case of, say, mining coal to generate electricity if the environmental impacts of mining and burning coal were not accounted for. Since these externalities of coal consumption extend into the future and the electricity generated is consumed almost immediately, there is no offsetting capital exchange. In this case, the environment is treated as a resource—consumed in exchange for wealth. Weak-side sustainability requires the environment to be treated as capital and not consumed as a resource, but it does allow the exchange of capital between its different forms.

Strong-side sustainability does not provide for shifting capital from one form to another; natural capital cannot be traded for an increase in real-estate equity, for example. Proponents of strong-side argument point to "critical natural capital" in the form of natural systems. We rely on natural systems to perform arrays of functions and services that are not provided for in the economy. Strong-side sustainability would require our valuation of nature to include not only the direct value of these services but also the systemic value. For example, the water treatment values of various wetlands have been calculated as ranging into the millions of dollars per year.[9] Others have calculated the value of wetlands in absorbing floods or acting as groundwater recharge zones. What is the systemic value of wetlands in a watershed? How do we account for the contribution that a given wetland makes toward maintaining biodiversity? Much of our attention, and therefore our regulatory approach, is focused toward managing the unique or the endangered, but what are the systemic values of the ordinary places and routine functions? When does the accumulated loss or degradation result in catastrophic failure of a system? Strong-side sustainability argues that "critical natural capital" cannot be equated to other forms of capital, since the valuation does not account for such systemic values.

So, our considerations may occur on a continuum anchored by these competing views of strong- and weak-side sustainability. This continuum of sustainability ranges from permitting actions that involve an exchange of capital between types (natural capital in exchange for economic capital for example) to prohibiting any exchange between forms of capital. The problems are obvious. If the cost of offsetting environmental impacts exceeds the economic value of a project, the project is not economically sustainable. If we account for the true environmental costs of a project and they exceed the economic benefit, how should the project be assessed? On the one hand, arguments come down to balancing cost models, but on the other, we are hamstrung from most undertaking any but the most simple economic activities. Most decisions under these rubrics will require thoughtful analysis to yield a sustainable outcome; even the idea of a sustainable outcome may differ widely from one perspective to another, from one place to another, and from one time to another.

Still, the designer's interest in sustainability is a practical one. Sustainability as an objective in design requires a deeper understanding and more measurable qualities than what a conceptual continuum will provide. Sustainability will require important changes in our thinking and values, as described by Thomas Berry,[10] Amory Lovins, William McDonough,[11] and many others. This is especially true of the thinking among certain professional groups.

Engineers, architects, landscape architects, product and industrial designers, and other design and planning professionals play important roles in the quality of our built environment and its impact on the environment and the future. Already, many individuals and firms have adopted a "greener" approach to their work. Still, development proceeds, even with green elements, with its impacts largely not accounted for. In the design processes in the United States, the design professional is but one element. Be it community officials and regulations, state and federal permits for particular actions, or regulations to adhere to, there are clients, standards, insurers, interest groups, and, at last, the consumer, all of whom participate in the process to a degree more or less. Then, there is the issue of what difference a single element in an individual project or an individual project itself can make on development as a whole. In the end, these might be difficult to quantify, but the overall impact of our decisions is measurable and what we have learned is that significant and fundamental change is in order.

Perhaps, a fundamental distinction must be made when thinking about sustainability. The word has taken on a vague gravitas that is used by everyone from philosophers to car salesmen, but it has no meaning beyond the context implied by the user. It seems important, but for the most part, we are not really sure what it means. It seems that there is a need that a book discussing the ethical implications of sustainability defines the term. The definition is even more critical from a designer's point of view, since the designer shoulders much of the responsibility for synthesizing the form that our intentions take.

What does it mean to say that design will lead to a sustainable outcome? The laws of nature dictate that anything we might conceive to manufacture or build will require the use of materials and energy. McDonough and Braungart have articulated compelling and workable principles that would lead to sustainable outcomes.[12] These involve a fundamental rethinking of how we evaluate and use materials and energy, how the things we design function, and how we define quality. In other words, these involve a rethinking of our values in a fundamental way, of how we might move from the current values and thinking about design to sustainable values and design. The discussion of sustainability among designers is broad. It encompasses the entire continuum of weak to strong sustainability.

Practicing designers have practical issues and concerns. Discussions of sustainability are replete with anecdotes that support or illustrate whatever point is trying to be made. In the end, what is needed, and what one hopes is at hand, is a fundamental shift in the values that underwrite the design process. This shift must occur in all stakeholders of the process, but since the design professionals are the core of the process, perhaps, this needs to happen with them most of all. Adaptation requires giving up or modifying the values of the past in favor of new values from which the paradigm of sustainability will take shape.

Before discussing adaptation, however, it is worth reviewing what obligations and expectation underpin the role and responsibility of the design and planning professionals today. It seems if we are to become a sustainable culture, our behaviors will be guided by values that inform our decisions and the norms that emerge from actions that we take in the years that lie before us. It requires us to act on what we believe to be true. In turn, this requires us to think about the values contained in our beliefs. Then, the word sustainability as it is currently used is insufficient. McDonough talks about abundance, growth, equity, and renewal.[13] Sustainability must be more than simply "green" outcomes if it is to be a social change. Green products and designs suggest that we can simply design our way to sustainability without any changes to ourselves and without any value shift. John R. Ehrenfield argues convincingly that a word that better expresses what is generally meant by sustainability is flourishing. Flourishing encompasses environmental, economic, social, and human well-being. If form is to be an expression of function, how does design express flourishing?

Why Sustainability Is Needed?

There is much to accomplish. Even discussing the issues has the tendency to devolve into recitations of apparently conflicting statistics. For example, between 1980 and 1995, per capita energy consumption in the United State fell, but the total energy consumption increased by 10% because of a 14% increase in population. From 1995 to 2005, the per capita trend in energy has been flat, perhaps even declining a bit, but as the population of the United States increases at a rate of about 3 million people per year, the total energy use rises as well. Likewise, while modern cars are 90% cleaner than cars from 1970, there are so many more of them that the efficiency gains are offset to some degree by the increase in the volume of pollution. Clearly, one can represent a variety of points by selecting which data to focus on and to present. Much of the public debate is characterized by incomplete data and half-truths that serve purposes other than open discussion and education.

The impacts of development and land use patterns have been well documented during the last 50 years. The environmental concerns are wide ranging and significant, but development is an important part of our economy and is necessary if the demands of a growing population are to be met. A 2007 University of Ohio study found that landscape fragmentation, the loss of cohesive patterns of connectedness in a landscape, had increased 60% from 1973 to 2000.[14] Fragmentation is known to be closely associated with a loss of biodiversity and resilience in landscapes. As development proceeds, leaving isolated "islands" of green space and planned narrow visual buffers to give some esthetic appeal to the finished site, they are commonly without significant ecological value. There are a variety of studies and reports that detail the public and personal health impacts of some development patterns. Human health impacts range from obesity, hypertension, respiratory problems, and even mental health concerns. The causes are equally diverse and include reduced air quality, traffic noise and vibration, sedentary lifestyle, and a loss of social capital.

There are higher public costs as well. Numerous studies have found that suburban development, as it is typically done, does not raise sufficient tax revenue to pay for itself and so drives the need for higher taxes for the existing communities. In 2003, a study found that for every new dollar of revenue raised by a new development, on average, $1.11 was spent in providing services to those same developments.[15] A report completed in Maryland found that school bus budgets in that state more than doubled to $492 million from 1992 to 2006. The miles driven by buses increased 25% in that time.[16] Although some states have laws that allow for impact fees, even a quick analysis reveals that these sorts of ancillary costs are usually not captured by them. In addition to these local impacts, we realize that human activities have significant impacts on global climate. People around the world have become more aware, and, perhaps, concern is being turned into action.

In 2007, Americans generated 254,000,000 tons of solid waste, from which about a third was recycled.[17] The average US household contributes 23 tons of greenhouse gases embodied in the consumer goods, foods, and services consumed every year.[18] The embodied energy in those products is difficult to calculate but represents a significant energy input. Since it is estimated that the greater percentage of consumer goods are used for less than 1 year and then disposed of, the energy spent in manufacturing, packaging, and shipping them is, in a sense, disposed of too. In addition to the degree of waste designed in to this system, there are the materials used.

This awareness is made more critical by the population increases expected in the coming decades. The United States currently has a population of more than 300,000,000 people, and it is expected to grow to 438,000,000 by 2050, an increase of nearly 3,000,000 people per year. It is expected that 20% of the US population in 2050 will be foreign-born legal residents and that 82% of the increase in population will be due to immigrants and their children and grandchildren. To respond to this population increase, it is necessary to build the equivalent of a city the size of Chicago every year going forward.

What will that development look like? Already about 80% of the buildings in the United States have been built since 1960. Buildings are responsible for 48% of the increase in greenhouse gases produced by the United States since 1990, an increase greater than emissions from either industry or transportation. A building constructed in the European Union typically uses about 25% of the energy of a similar building built in the United States. The patterns of growth in the United States have changed as well. A study completed by

Ohio State in 2007 found that today, most suburban development or sprawl is occurring in bands located from 55 to 80 miles from urban centers. This pattern of growth has been underwritten in part by road improvements that enable people to live further from the city centers and encourage more driving and more energy consumption.

A study of land development in the Washington, DC, area by Woods Hole Research Center found that development in the study area had increased 39% from 1986 to 2000. NASA concluded from its work that a 60% increase in total development should be expected in the DC metropolitan areas by 2030.[19] Given the environmental, economic, energy, and public health implications of development, as it has been done since 1960, there is a compelling argument that change is required. Still, we are facing significant population growth and with it demand for housing and services. It is expected that there will be an increase from a bit more than 310 million to about 438 million people in the US population by 2050, roughly a 35% increase or 3 million more people each year. The environmental costs of business as usual are difficult to imagine. Much of the population growth in the United States is occurring in the southwest and southeastern United States, the Sun Belt, where growing populations will put increasing pressure on already-strained supplies, and will require more development and more conservation.

Population growth alone, however, does not account for much of the development in the United States. Several studies have looked at the trends in growth and found that only about 50% of development in the United States can be explained by population growth.[20,21] What is worse is that states with growth control programs and legislation seem to fare no better than states without such controls, when it comes to limiting sprawl.[22] Development proceeds to a significant degree, not as a function of population but of public policy and practices that encourage expansion often at the expense of existing urban and suburban areas. Growth is subsidized in any number of ways, often at the expense of existing communities and development. New roads must be constructed, and sewers and water lines must be extended; with new housing comes the need for schools and community services. Very often, this growth is unaccompanied by real population growth, and more land is consumed to support the same number of people. If sustainability is the objective, these events might better be viewed as failures of planning rather than successes.

There were about 120 million residential buildings, 117,000 schools, and nearly 5 million office buildings in the United States in 2000. Each year before 2008, about 1.8 new residential buildings and 170,000 new commercial buildings were constructed (about 44,000 commercial buildings were razed each year).[23] As a whole, buildings accounted for 39.4% of the total US energy consumption in 2002, 67.9% of the total electricity use.[24] Runoff from the impermeable surfaces associated with roads, parking lots, and roof tops is the fourth leading source of pollution in rivers, third most important source of pollution in lakes, and second in estuaries.[25]

Studies and reports from around the United States continue to find disparities in housing markets, when viewed from racial and income bases. These disparities range from appraised values to sales as well as appreciation. As housing prices grew through the last few decades, even moderate housing grew out of reach for many low-medium- and low-income people. One such study looked at the minimum wage required to afford a two-bedroom apartment in 33 different markets in the United States. The researchers found that a worker earning the Federal minimum wage would have to work 108 hours a week to earn the rent for an average two-bedroom apartment.[26] Studies by the US Census Bureau show that housing affordability has consistently declined from the early 1980s through to 2007, even in the throes of the sub-prime mortgage era.[27]

Beyond these physical trends and their respective environmental costs lie equally important issues of human well-being. Human health issues in the developed world are often correlated to "lifestyles," which in turn are an outcome of various elements of design. A report produced by the Ontario College of Family Physicians found that the greater the population density of a community, the fewer the fatalities per 1000 people.[28] Areas with lower population densities have higher incidence of cardiovascular and lung disease, cancer, diabetes, obesity, and traffic injuries. The authors of the report recommend moving to a place with public transportation, which encourages walking and bike riding, and with adequate parks and open space, and engaging in the activities of the community. Most of these attributes are directly associated with design. In the United States through the first half or so of the twentieth century, lifespan increased to about 80 years but has remained fairly flat for the last 40 years or so. In the meantime, the length of time for which we live with disease or disability has increased dramatically and the cost of our health care has also increased dramatically. This suggests that we have built a system in which life is prolonged at ever-higher cost not only in terms of money but also in terms of well-being.

Still, all is not lost. The US Green Building Council has provided the Leadership in Energy and Environmental Design (LEED) certification to 550 buildings, with many more in the pipeline. Some builders and designers predict an increase of several thousand percent in green buildings, as demand surges among homebuyers and commercial real estate interests. Still, this represents a pitifully small number of buildings when compared with the total numbers. The trends indicate that the current practices of growth and development are not sustainable in any of the three key categories of the PCSD report. If we are to meet the challenge of building the equivalent of another Chicago every year and of the PCSD report, we must embrace a new approach to planning and design of the built environment.

Selling Sustainability

In the discussions among various professionals, a common concern rises from time to time. For purposes of this examination at least, it seems for the most part that design professionals engaged in the discussion tend to fall into one of two broad camps when it comes to issues of sustainable development or environmental considerations in general. The first group is concerned with compliance only and is sure to account for environmental impacts only to the degree necessary for compliance. The other group openly trades on its commitment to environmental quality and to minimizing the impact of development. It is important to note that neither group necessarily has more knowledge or even a greater interest in environmental quality. Rather, it seems that circumstance, or what might be called practice inertia, might have a greater influence than any cogent intent, limited knowledge, or disinterest.

There is awareness among all of the professions that environmental problems exist and that environmental quality has suffered because of development practices in the past. There is even fairly broad agreement among them that these practices need to change. Disagreement appears to lie in where the responsibility for change lies, just what does change imply, and, more to the point, the degree to which the design professional is accountable for either. Firms falling in the first group, which I will call the compliance group, believe strongly that their professional duty is met by adhering to the practices of the past and adapting, as required, to changes in public policy. In essence, it is the responsibility of regulatory agencies and the public change rather than of professionals to direct change. When probed, the most frequent concern about acting in a leadership role among the compliance group is the loss of business and, to a lesser degree, lack of a consensus of what sustainable design even is. The argument is that if a firm should do more to protect the environment, it will inevitably cost the client more and the client will take its business somewhere else. In these circumstances, the firm is competing primarily on the basis of the cost of building the design product. The pragmatic designer/manager in a design firm is concerned with the immediacy of meeting payroll, overhead, and other practical concerns. The environment can seem somehow more esoteric than these consequential concerns; economic realities trump environmental realities.

There may also be resistance among many more senior professionals to new methods and materials, perhaps because successful professionals are promoted within organization or have achieved their professional and business success largely on the basis of their technical skill. As managers, however, they may no longer "practice" their profession at the same intensity, tending to oversee younger professionals or groups rather than spending actual time "in the weeds," solving design problems. They reach positions of authority and status based on a specific frame of technical knowledge and experience, and they tend to attempt to preserve the circumstances of their success. That is to say, they resist changing to methods and practices with which they are unfamiliar and with which they have no first-hand success. As a result, the more entrenched and experienced the management is, less likely it may be to be truly innovative or adapt to change, unless of course those are the very traits and skills on which the firm trades.

On the other hand, there are those professionals and firms that embrace change and take strong leadership positions in the marketplace, promoting new methods and approaches. I call these the "boutique" firms, because it is change that is at the heart of the firm and environmental quality may be the medium in which they work. These individuals and firms tend to be younger, though within them there may be very experienced and older professionals. The key is that, to distinguish themselves in the marketplace and to overcome the inertia of the more established firms, they must offer something different than the older competitor, and environmental quality is part of their commitment. These firms tend to be newer, smaller, and much flatter in terms of hierarchy; however, it would be a mistake not to recognize that a good deal innovation comes from very large, well-established firms. Their work is distinguished from the "compliance" firms by virtue of innovation and uses of new materials or familiar materials in new ways. They compete on the basis of innovation and solutions rather than price.

Clearly, these are generalizations, and therefore, they necessarily fail to describe the variety of actual firms and circumstances in the marketplace. But while there are shortcomings in the descriptions, they serve to frame a common dilemma of the working design professional and the design firm. In this marketplace, both types of firms are faced with a changing marketplace, where mere compliance may not be sufficient and where there are attempts to regulate innovation. In both cases, the design professional brings value to the client and the project, though the quality of the value may be quite different. Data show that many of the sustainability practices have actually no greater costs and, in some cases, lower life cycle costs than the alternative, but these demonstrations rarely sway the compliance-oriented professional.[29] Firms and professionals tend to do what they have done in the past, that is, they attempt to repeat their successes over and over in a sort of practice inertia.

The design professionals, whether in the private or public sector, is unique by virtue of their knowledge and the professional status it earns them. There have been a variety of suggestions for how this role should be characterized. Some have suggested that because of the scope of their influence, design and planning professionals should go so far as to adopt an oath similar to physicians—a design profession version of the Hippocratic Oath. Others have argued that the process is beyond the influence of the individual practitioner, that the responsibility for development impacts lie with the stakeholders, and that design professionals are constrained by the process.

As this is being written, at least 8 states and 113 cities have adopted some version of "green building" codes. Some governments have used the LEED process to guide their actions. Other groups are continually developing specific approaches to guide their member; for example, it seems each new iteration of the American Society of Heating, Refrigeration and Air Conditioning Engineers (ASHRAE) includes updates and fine tuning of their standards. Municipalities reflect the interests of their citizens and are beginning to have a greater interest in green buildings, increased energy efficiency, and reduced environmental impact. Design professionals have played important roles in creating the awareness and providing the examples that have inspired others; however, the overwhelming amount of design and development evidenced in the marketplace is firmly rooted in the twentieth century.

Design professionals have the opportunity to provide clear leadership through example in this era of change. In the end, they will lead and guide their clients to the future, or they will be led through public policy and the innovators in their midst. There is a degree of uncertainty in the marketplace of design, because the comfort zone is being stressed by new standards of quality. The individual designer and the firm must consider the implications of following instead of leading the way.

Among the problems of sustainable design is also its greatest opportunity for the professional. It is common to hear professionals complain that sustainable design does not offer the standards on which to guide their efforts. This is true in many cases. To some extent, it is a reflection of how new the efforts are. To another extent, it is a reflection that sustainable design is likely to differ in important ways from project to project, from place to place, and even from time to time, as we learn more. What designers risk are "standards" being developed that make work more predictable but limit true sustainable design, constrain creativity, and stifle meaningful problem solving. Sustainable design "standards" may produce "greener" projects but not true sustainability. Sustainability will not be achieved by creating and merely following standards. A single green project, or collection of separate green projects, or products will not result in a sustainable society and economy. In the words of William McDonough, "being less bad is not the same as being good."[30] It is unlikely true that sustainability will be produced by prescriptive "standards" produced in a legislative process. While standards that contribute to maintaining a sustainable state of affairs may someday emerge, producing sustainability will initially require outcomes-focused or performance-based standards.

Sustainability is ultimately systemic; it is in the relationships between things, not in the things themselves. Standards imply and often result in design without analysis, using a single solution in place of all possible solutions. For standards to work in creating sustainability, we would require that there be universal parameters available to define all the relationships that we find in a complex environment. Sustainable design requires thoughtful analysis, a deeper understanding of the implications our work has on the relationships between things in our world. The opportunity to lead that process belongs to the design community. The nature of sustainable design is likely to produce a more collaborative process than has been the experience of most practicing design professionals. The collaboration is likely to include new professions and stakeholders as well. Clearly, to reflect the values of a community or a marketplace, the designer must reach out and understand the values that underlie flourishing. To lead that process requires the design professions to develop a clear vision of their role and the desired outcome. To have the credibility to "sell" their place at the center of these developments, they must commit themselves to understanding the broad implications of sustainability and its success. This book is about the ethical implications of that role.

Endnotes

1. Anderson, Albert A., "Why prometheus suffers: Technology and the ecological crises." *Society for Philosophy and Technology*, 1(2), 1995.
2. Anderson, Albert A., "Protagoras," Agora Publications, Baltimore, MD, 2008.
3. President's Council on Sustainable Development, May, 1999, http://clinton2.nara.gov/PCSD/Publications/TF_Reports/amer-top.html
4. CEF, "Is sustainability sustainable in the age of trump?" April 3, 2017 http://www.corporateecoforum.com/sustainability-sustainable-age-trump/, Accessed May 20, 2018.
5. Greenbiz, State of Green Business, https://www.greenbiz.com/microsite/state-green-business, Accessed May 20, 2018.
6. Maslow.com, http://www.abraham-maslow.com/m_motivation/Hierarchy_of_Needs.asp, June 26, 2009.
7. Convergence 12, University of California at Santa Barbra, http://convergence.ucsb.edu/article/fisheries-salvation, Fall 2008.
8. Malthus, Thomas, "Essay on population," http://www.fordham.edu/halsall/mod/1798malthus.html
9. Turner, Marjut H., and Richard Gannon, NCSU Water Quality Group, "Information on Wetlands" North Carolina State University, http://www.water.ncsu.edu/watershedss/info/wetlands/values.html, 2000.
10. Berry, Thomas, "Ethics and ecology" a paper delivered to the Harvard Seminar on Environmental Values, Harvard University, www.ecoethics.net/ops/eth&ecol.htm, April 9, 1996.
11. McDonough, William, "A centennial sermon: Design, ecology, ethics, and the making of things," in *Ethical and Environmental Challenges to Engineering*, Gorman, Michael E., Matthew M. Mehalik, Patricia Werhane, editor, Upper Saddle River, NJ, Prentice Hall, 2000, pp. 90–97.
12. McDonough, Willam and Michael Braungart, *Cradle to Cradle: Remaking the Way We Make Things*, New York, NY, North Point Press, 2002.
13. McDonough, William, Remarks at 2006 TED, http://www.youtube.com/watch?v=IoRjz8iTVoo.
14. Irwin, Elena G. and Nancy E. Bockstael, "The evolution of urban sprawl: Evidence of spatial heterogeneity and increasing land fragmentation." *Proceedings of the National Academy of Sciences (PNAS)*. 104 (52), 20672–20677.
15. Goetz, Stephan Juergen, James S. Shortle, John Clark Bergstrom, "*Land Use Problems and Conflicts*" *Causes Consequences and Solutions*. Routledge, Taylor and Francis Group 2004.

[16] Sewell, Chris, Ahern, Lena-Kate and Hartless, Allison, "Yellow school house blues" 1000 Friends of Maryland, October 2007 http://www.friendsofmd.org/reports.aspx

[17] Municipal solid waste generation, recycling, and disposal in the United States: Facts and figures for 2007 http://www.epa.gov/osw/nonhaz/municipal/pubs/msw07-fs.pdf

[18] "Consumer-oriented life cycle assessment of food, goods and services," Christopher M. Jones, *University of California, Berkeley* Daniel M. Kammen, *University of California, Berkeley* Daniel T. McGrath, *University of California, Berkeley* http://repositories.cdlib.org/bie/energyclimate/jones_kammen_mcgrath_030308/

[19] Woods Hole Research Center, "Land use change and the chesapeake bay ecosystem" http://www.whrc.org/midatlantic/modeling_change/SLEUTH/maryland_2030/md_scenarios.htm

[20] Pendall, Rolf, "Sprawl without growth: The upstate paradox" Brookings.edu http://www.brookings.edu/reports/2003/10demographics_pendall.aspx, October 2003.

[21] Kolankiewicz, Leon and Roy Beck, "100 Largest Cities," http://www.sprawlcity.org/studyUSA/index.html, 2007.

[22] Anthony, Jerry, "Do state growth management regulations reduce sprawl?" *Urban Affairs Review*, 39(3), 376–397, 2004.

[23] USEPA Green Building Workgroup, *"Building and the Environment: A Statistical Summary,"* Environmental Protection Agency Green Building, Washington, DC. Available at: www.epa.gov/greenbuilding/pubs/gbstats.pdf (accessed October 2, 2008), 2004.

[24] Annual energy Review 2003, DOE/EIA-0384 Energy information administration, US Department of Energy Sept. 2003.

[25] The National Water Quality inventory: 2000 Report to Congress. USEPA 2000.

[26] "Out of reach: America's growing wage rent disparity," National Low Income Housing Coalition Annual Report,

[27] Savage, Howard A., "Who could afford to buy a home in 2002?" Current Housing Reports US Census Bureau, H121/07-1 July 2007.

[28] Taekmoto, Neil, cooltownstudios.com October 31, 2005 http://www.cooltownstudios.com/2005/10/31/city-living-healthier-than-in-the-suburbs.

[29] Winer-Skonovd, Rebecca, Dave Hirschman, Hye Yeong Kwon, and Chris Swann, *"Synthesis of Existing Cost Information for LID vs. Conventional Practices"* Center for Watershed Protection, Ellicott City, MD. 2006. http://www.mpcnaturalresources.org/PDF/StormWater-PDF2009/LID%20vs%20Conven%20Costs%20Memo.pdf

[30] McDonough, William "The next industrial revolution."

2

Professional Ethics

We are what we repeatedly do. Excellence, then, is not an act, but a habit.

Aristotle

What Is a Profession?

What is it about certain classes of work that they are distinguished as "professions"? It is common today to refer to almost all sorts of employment as "professional"; from trades people to neurosurgeons, is there no distinction? Clearly, some forms of work can be distinguished from others, and they merit the claim as "professional." One clear measure of professionalism is public sanction through licensure. Ostensibly, the reason for licensure of a particular skill or set of skills by a public authority is to protect the public health and safety and the environment. The government has a duty to protect citizens from harm.[1,2] Licensure requires the candidate to meet some requirements, demonstrating some benchmark of a minimal degree of competence in the skills required to practice the form of work being tested. In some states and some cases, licensure may be given provided that the candidate provides a demonstration of acceptable past work or licensure in another state. In either case, it is assumed that the licensed professional meets the threshold of minimal competence to practice the specific occupation.

These controls are encroachments in the free market in that they prevent those who might otherwise be interested in participating in the work restricted to the licensed groups. While it is true that the marketplace would likely correct incompetent performance by taking its business elsewhere or through injunction, both of these are remedies after the harm has occurred. If the government has the duty of protection, it is obliged to prevent foreseeable harm. We would all agree that it is better to have our bridges designed by individuals who have demonstrated their competence before their first bridge was underway.

Having lumped all occupational licenses together, it is necessary to distinguish between professional licenses or registration and occupational licenses. Occupational licenses are earned through an examination or assessment process that is organized and administered by the state, whereas professional licenses and registrations are attained from examinations organized and administered by the professions themselves and are accepted by the state. The professional license usually requires extensive education and experience before a practitioner may sit for the exams. Licensing, whether professional or occupational, is intended to exclude unqualified people from practice. Licensure of the professions is appropriate, however, because of the special knowledge and potential for harm by unqualified practitioners and can be justified within a capitalist system only with regard to the government obligation to protect its citizens. To some extent, the predictable flaws in the system might be added to the privileges claimed by the professions of having first attained the skills and knowledge and then the admission to the profession.

While the scope of professions is broad, they share several commonalities. First, they are all services in one form or another; they usually involve an issue of personal service, public safety, or environmental protection, where the potential for harm through incompetence is significant. In general, licensure laws are passed because of experience with the harm caused by unqualified practitioners and for the encouragement of presumably qualified members. The state's interest in certifying qualified persons is so that the public has a measure to gauge those who have demonstrated competence from those who have not. If private organizations alone determine the methods of certifying their members to be competent and ultimately sit in judgment of that competence, can the public rely on the organization to be primarily concerned with the interests of the public? In the United States, Certified Public Accountants are certified by a private board rather than a state-sanctioned license or registration, but it serves as an example of the risks of a profession overseeing itself. In the early 1990s, the private board changed its rules pertaining to conflict of interest to allow firms auditing companies to also be able to act as consultants to those same clients. Many believe it is this change that led to the problems of Enron and the demise of Arthur Anderson. More to the point, was the public's interest served in this case of nonpublic certification? The collapse of banks and the follow-on economic crisis beginning in 2007 have been described ultimately as the result unregulated self-interest among financial professionals. The reaction to this failure to regulate itself is that the financial industry is likely to end up with more government regulation. The failure of professions to police themselves creates a compelling interest for sanction through the state.

Criticism of licensure makes the argument that licensure serves to only restrict some that would otherwise pursue the occupation and to allow those who are licensed to enjoy a monopoly and conceivably higher prices. Such restriction on the freedom of choice in selecting an occupation or choosing from whom to purchase services is unfair in the view of some. States choose to license a broad array of occupations, from cosmetologists to physicians and from well diggers to structural engineers.[3] Then, these observations prompt the following questions: Is occupational licensure consistent with the basic principles and values of capitalism? Is it a violation of the free-market ideal? Does it restrict individual freedom of a person to freely choose a career?[4] Licensure is clearly a restriction on a person's choice of occupation but only to the extent that it requires them to demonstrate the minimal level of competence.

This approach does raise concerns though. Actual processes of certification, however, are usually left largely in the hands of the profession to determine what minimal competence for licensure is and how it is to be earned and measured. While in theory this may sound fair, in practice, it maybe even more restrictive than the licensure laws. It may be true, for example, that some professions have used the examination process to limit the number of practitioners or to encourage more practitioners by increasing or decreasing the difficulty of the examination or by changing the eligibility requirements to take the exam. In essence, the members of the profession determine to raise or lower the bar of minimal level of competence based in the interests of the profession. This manipulation is largely for the benefit of the profession, as opposed to the benefit or protection of the public. To the extent that these practices are not consistent with the protection of the public and the environment, they deserve to be criticized. On the other hand, if professions were not sanctioned by the state, as private organizations, they would have no authority to restrict unqualified or incompetent people from practicing. The public's interests, however, clearly outweigh these concerns and compel licensure sanctioned by the state. Passing laws or even regulations that define a minimal level of competence would risk becoming outdated quickly, being subject to the political process rather than technical concerns best understood by qualified practitioners, and ultimately undermining the very qualities of competence they mean to assure. The nature of special knowledge that forms the foundation of the profession is necessarily always adapting to the world as new concerns, style, materials, and needs arise. The definition of minimal competence is necessarily a dynamic and moving target. A demonstration of minimal competence is required to enter the profession, but the various states wisely leave the determination and evaluation of an individual's competence to the profession itself.

All professions also share a common thread in their relationship with society in the form of some distinctive knowledge earned as a result of study and training. In its most basic form, a profession can be defined by five characteristics that "…all professions seem to possess: (1) systematic theory, (2) authority, (3) community sanction, (4) ethical codes, and (5) a culture."[5] A systematic body of theory or specialized knowledge supported by a body of theory requires ongoing rational development of the theory. In turn, this results in professions having both theoreticians (research) and practicing professionals. In a professional relationship, the client relies on the professional to determine what must be done and what is best for the circumstances and for the client. This professional authority is distinct from other relationships wherein the consumer or customer knows what is wanted and shops around to find it.

In the relationship between the professional and client, the client seeks the professional's special knowledge, informed opinion, and recommendations; there is an inherent inequality in the relationship. This distinct knowledge and professional authority lead to community sanction or approval of the profession's scope of authority or power. These powers include control over the training centers of the profession (certification or accreditation of educational institutions), control over admission to profession (administration of examinations), and a confidential relationship with client. Professionals are further distinguished from trades by the adoption of a code of ethics usually by a professional association or peak organization. Finally, professions are socially distinguished through the development of values, norms, and symbols of a professional culture.

The Engineers Council for Professional Development (ECPD), as reported by Firmage, describes the attributes of being a professional as:

1. "They must have a service motive, sharing their advances in knowledge, guarding their professional integrity and ideals, and rendering gratuitous public service in addition to that engaged by clients.
2. They must recognize their obligations to society and to other practitioners by living up to established and accepted codes of conduct.
3. They must assume relations of confidence and accept individual responsibility.
4. They should be members of professional groups and they should carry their part of the responsibility of advancing professional knowledge, ideals, and practice."[6]

Society sanctions the profession because of the perceived benefit of the profession's activities and the desirability of the outcomes of those activities. In exchange for the sanction, society expects a certain level of performance from practitioners, a degree of what is generally described as "professionalism." Professionalism in this sense refers not only to the special knowledge that members of a profession are thought to have but also to the duties and responsibilities that the individual practitioner has toward society.[7] It might be said that it is simply by virtue of having any special knowledge or skills that an individual would also have these special obligations. Every individual has a right and duty to pursue his or her own well-being but only to the extent that such acts do not harm to another. Further, every person has an ethical obligation to make some sacrifice to prevent harm to another person or to assist a person in need. Professionals, however, have developed special skills and are afforded social advantage because of them and therefore acquire and accept a greater responsibility than those who have not.[8,9] Not unlike the Promethean example, designers have special training and knowledge that allow them to better anticipate and assess the effects of actions than nonprofessionals, and they must be expected to use these skills. They are expected to use the forethought provided by special training and skills and the confidence sanctioned in public licensure.

"Due Diligence" in Codes of Ethics

Codes of ethics are a common means of corporate or professional self-regulation. The advantages of self-regulation given by business over government intervention usually fall into one or more of the following characterizations. The specialized knowledge of those within a business or professional to understand the unique character and practices in question is a key element in the self-regulation argument. The second basis for preferring self-regulation extrapolates that expertise into the ability of peers to pressure colleagues to abide by preferred practices or behavior. Finally, self-regulation is also motivated by avoiding costs that businesses commonly associate with government regulation.[10] Recent events, however, have demonstrated the inherent weaknesses in self-regulation.[11]

Ian Maitland criticizes business's "ability to regulate itself as an outcome of manager's single-minded preoccupation with profits maximization."[12] He reports that though corporations benefit from acts performed in the public interest, they collectively have that benefit, regardless of whether or not they contribute it. For example, a corporation benefits from having a cleaner environment, no matter if it actually contributes to it or not. Since corporations tend to act as rational economic persons, that is, in their own self-interest, they would tend to accept the benefit of the public goods made available by common action but would avoid the cost of contributing (e.g., lower costs of production by avoiding pollution controls), the so-called "free rider" problem.

Maitland writes that self-regulation is likely to work best in the presence of a "peak organization" that is a large group that encompasses many businesses or individual professionals. The "more inclusive or encompassing the organization, the larger the fraction of society it represents...the higher the probability that it will oppose self-serving behavior (by the sections of its membership) that inflicts its external costs on the rest of society."[13] This view of the role of the peak organization raises several questions, particularly in the United States, where a cultural of individuality and the political nature of regulation serve to minimize the effectiveness of peak organizations.[14] Also, the nature of the professions is designed to be exclusive, limiting full membership largely to qualified professionals and thus limiting the fraction of the population Maitland speaks of to a narrowly defined group with deeply shared interests. Despite these assertions, there is significant public reliance on and reference to the codes of ethics of professionals established by peak organizations in the United States.

Assessing Due Diligence

In 1991, the United States promulgated new sentencing guidelines for federal courts. A portion of these guidelines addressed the sentencing of organizations convicted of a federal crime. In the past, corporations found guilty of a crime were often given punishment viewed as too lenient, with the problem being how to punish a corporation. Often, there was no person with criminal liability, and to force the corporation out of business also hurts employees and communities.[15] The new guidelines established a means of determining levels of offense and provided for more punishment for offenses and behavior found to be more egregious. Also contained within the guidelines was a process of evaluating corporate behavior by considering whether the corporation had in place a program to discourage unwanted actions by employees and encourage practices that it found to be ethical and acceptable. This has come to be known as the due diligence test of corporate ethics programs.[16] "Due diligence" is determined by a seven-part test[17]:

1. The existence of a set of standards and procedures that would be expected to reduce the criminal conduct of a corporation's employees or agents
2. A responsible person or persons within the corporation who oversee compliance
3. A practice of avoiding allowing discretion to persons that the corporation knew or should have known might engage in illegal activities
4. An effective program of communicating standards and procedures to employees and agents
5. Committing to achieving compliance with the standards and procedures through action
6. The enforcement of the standards and procedures through appropriate disciplinary action
7. Demonstration that, once detected, steps are taken to respond to the offense and to avoid its reoccurrence

As a general guideline, the seven-step due diligence test may provide some insight into the ethics programs of peak organizations. Though admittedly not a precise fit, it does accurately portray at least some of the relative strengths and weaknesses, and such comparisons could be useful tools for the public in evaluating the conduct and performance of design professionals.

The areas of greatest concern in this analysis were as follows:

- The general lack of awareness or recognition of stakeholder in the design process and the professional's obligations to them. Although there is a general acknowledgment of the professional's obligation to protect the public health and safety, the terms of the protection could be seen as paternal rather than collaborative. None of the codes reviewed for this exercise contained any language about stakeholders beyond the professionals' duties to the public welfare, the client, the law, and the profession itself. Although there were references to the professional's obligations in civic affairs, these seemed to be in the context of advancing the interests of the profession through indirect action, rather than any education or service objective.[18]

- The inability of the peak organization to disseminate information or training pertaining to its code of ethics. Each organization requires members to acknowledge the code of ethics at the time of joining or renewing their membership by signing a statement to that effect. All other contacts require an individual to seek the information, and this tends to occur only when the individual is confronted by a problem.[19] Dissemination of information is problematic, since compulsory training would likely not be welcomed by members. Still, several of the associations reviewed maintained visible and active links to their code of ethics and provided resources for members and ongoing communications through regular channels.

- The lack of enforcement capacity in some cases. Enforcement of the codes seemed to vary a great deal. Most of the organizations contacted for this book elected not to respond directly to questions regarding enforcement.

Table 2.1 is a straight-ahead comparison of the seven steps of the due diligence test compared with the four selected codes of professional ethics. It is important to note, for example, that some of the seven tests are not suited for precise application to a peak organization, such as those selected, and some allowance must be made in this analysis. For example, the third step in the process requires an organization to avoid discretionary authority to persons that the corporation knew or should have known might engage in illegal activities. Professional organizations have no such power or authority, except in the determination of who is allowed to become a member. All states that require licensure have a restriction that disallows felons and requires two licensed professionals to certify as to the character of the applicant. Professional organizations usually require members to be licensed professionals. The process would seem to respond to the spirit, if not the letter, of step 3. The seven steps may be an effective, though imperfect, general guide to the effectiveness of codes of professional ethics by virtue of measuring the due diligence of the peak organization.

The American Society of Civil Engineers (ASCE) and the American Society of Landscape Architects (ASLA) provided general data regarding the number of cases and enforcement for preceding years. In general, the experience of both groups seems similar. ASLA and ASCE each reported about 5 to 7 cases per year for offences ranging from plagiarism to issues relating to qualifications and fiduciary matters. Actions ranged from expulsion from the organization to publication of the offender's name and lesser actions. It is difficult to assess what enforcement looks like in other professional groups;

TABLE 2.1

Comparison of Codes of Ethics from Selected Peak Organizations for Design Professions

Element of Published Code of Ethics (Corresponding Element of 7-Part Test, 24)	American Society of Landscape Architects (ASLA)	American Institute of Architects (AIA)	American Society of Civil Engineers (ASCE)	American Consulting Engineers Council (ACEC)	Industrial Designers Society of America (IDSA)	American Association of Mechanical Engineers (ASME)	American Institute of Certified Planners (AICP)
Protection of public health and safety (1,2,3,5)	Yes	Yes	Yes	Yes	Yes	Yes	Yes
Protection of environment (1)	Yes	Yes	Yes	No	Yes	No	Yes
Recognition of human rights (1,5)	Not specifically	Yes	Not specifically	No	Yes	No	Yes
Protection of cultural heritage (1,5)	Yes	Yes	No	No	No	No	Yes
Nondiscrimination (1)	Not specifically	Yes	No	No	Yes	No	Yes
Prohibition of unfair competition (1,3,5,6)	Yes	Yes	"Build…reputation on merits of their services and shall not compete unfairly with others."	Yes	Yes	Yes	Yes
Limit practice to areas of competence (3,4,5,6)	Yes	Yes	Yes "…perform services only in areas of their competence."	Yes	Yes	Yes	Yes
Standard of care described (4,5,6)	Yes	Yes	Yes	Vague	Yes	Yes	Yes

(Continued)

TABLE 2.1 (Continued)

Comparison of Codes of Ethics from Selected Peak Organizations for Design Professions

Element of Published Code of Ethics (Corresponding Element of 7-Part Test, 24)	American Society of Landscape Architects (ASLA)	American Institute of Architects (AIA)	American Society of Civil Engineers (ASCE)	American Consulting Engineers Council (ACEC)	Industrial Designers Society of America (IDSA)	American Association of Mechanical Engineers (ASME)	American Institute of Certified Planners (AICP)
Limit practice to areas where judgment is not compromised (4,5)	Yes	Yes	Yes	Yes	Yes	Yes	Yes
Recognition of other professions	Yes	Yes	No	No	No	No	Yes
Avoid conflicts of interest (3,4,5,6)	Yes	Yes	Yes	No	Yes	Yes	Yes
Disclosure for public safety (1)	Yes	No	Yes	No	No	No	No
Disclosure for environmental issues (1)	Yes	No	Yes	No	No	No	No
Full disclosure (for fees accepted for public statements, to clients, and others who rely on advice, during proposal/solicitation for work. Any conflict of interest	Yes	Yes	Yes	No	Yes	No	Yes

(Continued)

TABLE 2.1 (Continued)

Comparison of Codes of Ethics from Selected Peak Organizations for Design Professions

Element of Published Code of Ethics (Corresponding Element of 7-Part Test, 24)	American Society of Landscape Architects (ASLA)	American Institute of Architects (AIA)	American Society of Civil Engineers (ASCE)	American Consulting Engineers Council (ACEC)	Industrial Designers Society of America (IDSA)	American Association of Mechanical Engineers (ASME)	American Institute of Certified Planners (AICP)
Professional development	Yes	Yes	Yes	Yes	Yes	Yes	Yes
Limits on advertising	"...shall not mislead..."	National Ethics Council, appointed by AIA board	"...in a way does not contain misleading language or is in any other manner derogatory to... the profession..."		No	No	No
Protect interests of, advocate of client's interest, confidentiality	Yes	Yes	Yes	No	Yes	Yes	Yes
Seek to raise standards of design excellence	Yes	Yes	No	No	Yes	Yes	Yes
Seek opportunities for public service	Yes	Yes	"...seek opportunities to be of constructive service in civic affairs..."	No	Yes	No	Yes
Duty to report ethics violations (6,7)	Yes	No	No	No	No	No	No

(Continued)

TABLE 2.1 (Continued)

Comparison of Codes of Ethics from Selected Peak Organizations for Design Professions

Element of Published Code of Ethics (Corresponding Element of 7-Part Test, 24)	American Society of Landscape Architects (ASLA)	American Institute of Architects (AIA)	American Society of Civil Engineers (ASCE)	American Consulting Engineers Council (ACEC)	Industrial Designers Society of America (IDSA)	American Association of Mechanical Engineers (ASME)	American Institute of Certified Planners (AICP)
Authority to hear ethics charges (2,6,7)	Yes	Yes	Yes	None listed	None listed	Yes	Yes
Authority to discipline (2,5,6,7)	Letter of admonition, letter of censure, probationary suspension, termination	Admonition, censure, suspension, termination of membership	None listed	None listed	None listed	Yes	Yes

however, an unscientific survey of professionals known to me identified only one person with firsthand knowledge of an ethics action taken by a professional organization and that was more than 20 years ago and had to do with fraud (falsification of education). Every person questioned had firsthand knowledge of an act by a peer or colleague that they defined as unethical, but no one felt a responsibility to act on this knowledge. Assuming that these experiences have some broader application, it would seem that unethical behavior is brought to the attention of the peak organization infrequently and there is some degree of tolerance for unethical behavior among colleagues. These present particular problems for the public in its ability to rely on the professionals to police themselves and for the interests of the public to be considered in any but a fairly paternal way. On the other hand, the experience and reporting of ASLA and ASCE suggest that when a complaint is made, the professions respond directly and, where deemed appropriate, forcefully. Even in the absence of complaint and enforcement data, the depth and balance contained in some of the codes made them useful and practical tools for the professional and would seem in principle to serve the interests of some stakeholders well. Other organizations have rather anemic process and either uncertain policies or a policy restricting public access to the information.

Evaluating Codes of Ethics

These seven steps become a de facto standard by which corporate ethics programs are measured, but what of the ethical codes of peak organizations? It is true that peak organizations of professional groups are not strictly corporations in the sense that the guidelines were conceived and unlikely that a professional group could ever be prosecuted in the same manner as business entities have been, but the public relies on the quality of the professions and, by reference, its ethical canon. In fairness, the codes of professional ethics reviewed are not corporate programs; however, holding them to a due diligence test may be instructive. The interest here is focused on the general form and content of the published codes of ethics for selected peak organizations of design professions[20] and how they might measure up against the due diligence test.

Table 2.2 is a compilation of different elements from the published code of professional ethics of ASLA, the American Institute of Architects (AIA), ASCE, and the American Consulting Engineers Council (ACEC).[21]

All of the codes of ethics reviewed share a general concern with the public health and safety, much in terms of the way of business protocols such as competition based on professional merit, professional development, some discussion of standard of care, avoidance of conflicts of interest, and the

TABLE 2.2
Comparison of Selected Codes of Professional Ethics to Due Diligence Test

Element	American Society of Landscape Architects	American Society of Civil Engineers	American Institute of Architects	American Consulting Engineering Counsel	Industrial Designers Society of America	American Institute of Certified Planners
The existence of a set of standards and procedures that would be expected to reduce the unethical conduct	Yes	Yes	Yes	Yes	Yes	Yes
A responsible person or persons within the corporation that oversee compliance	Yes (chairman of Ethics Committee)	Yes (members of National Ethics Council)	Yes (National Ethics Council)	No	Yes	Yes
A practice of avoiding allowing discretion to persons that the corporation knew or should have known might engage in illegal activities	Yes (by virtue of licensing restrictions)	Yes (by virtue of licensing restrictions)	Yes (by virtue of licensing restrictions)	N/A ACEC does not have individual members, only firms	Yes	Yes

(Continued)

TABLE 2.2 (Continued)
Comparison of Selected Codes of Professional Ethics to Due Diligence Test

Element	American Society of Landscape Architects	American Society of Civil Engineers	American Institute of Architects	American Consulting Engineering Counsel	Industrial Designers Society of America	American Institute of Certified Planners
An effective program of communicating standards and procedures to employees and agents	Yes	Yes	Yes	No	Difficult to assess	Yes
Committed to achieving compliance with the standards and procedures through action	Active enforcement of rules, communication to members	Yes. Publications, presence on website, links to resources and help line, active enforcement of rules	No response	No response	No response	Yes
The enforcement of the standards and procedures through appropriate disciplinary action	Process in place, range of sanctions possible	Process in Place, range of sanctions possible	Process in place, range of sanctions possible	No	Process in place	Yes
Demonstration that steps are taken to respond to the offense and to avoid its reoccurrence	Yes	Yes	No response	No response	No response	No response

maintenance of client confidences (see Table 2.1). The protection of public health and safety is expected because each of these professions requires a license or registration in the state in which it is performed, and the basis of the authorizing legislation for such a license is the protection of the very same public health and safety. In general terms, the codes are similar; however, the comparison does offer insight into areas of dissemination and enforcement, in particular.

Most of the codes reviewed are dedicated to the professional's conduct in business matters regarding to relations with clients or professional peers. ASLA and AIA include statements regarding the role of other professionals and the duty to recognize the contributions of other professionals. Beyond the general statements regarding public health and safety, references to the environment, and limiting one's work to areas in which one is competent, there is little in any of the codes that deals with a duty of a design professional to a larger group of stakeholders.[22] The codes would appear to set up a potential conflict between the interests of the client and the interests of other stakeholders. The ASCE code of ethics appears to require a preference to the interests of the client,[23] whereas the ASLA code includes a qualification or limitation on the duty to a client. All organizations' codes include some reference to the confidentiality of information acquired in the course of professional activities, as would be expected. ASLA qualifies this burden to maintain confidence of information learned "in the course of their professional activities which they have been asked to maintain in confidence, or which could affect the interests of another adversely. *Unique exceptions: To stop an act which causes harm; a significant risk to the public health, safety and welfare, which cannot otherwise be prevented; to establish claims or defense on behalf of members; or in order to comply with applicable law, regulations or this Code*"[24] (emphasis added). There are several instances in the ASLA and AIA codes of ethics where a statement is clarified as to the duty or limits of a duty of the professional. These notes, such as the one noted earlier, recognize the possible conflicts that might occur between obligations and provide important guidance to the professional in his effort to comply with the code.

ACEC and ASCE, for example, have identified a responsible party or a process for addressing ethical violations as part of the Codes (see Table 2.1), though both organizations have a process and ASCE does refer to the process in the introduction of its code of ethics. Whether the absence of an enforcement format as part of the code of ethics weakens the code is difficult to assess; however, it does muddle the appearance of effectiveness. If the public is to rely on the various professional groups to police themselves, it is important to assess their ability to do that. On the other hand, ASLA and AIA have identified responsibility, contacts, and an outline of the process.

ASLA and ASCE go on to inform members of an obligation to identify members that do not comply with the codes of professional ethics. These raise the questions as to whether such obligations are embraced by members and, if not, whether there actually is enforcement. It could be difficult for an organization to maintain membership if its members were concerned about constant surveillance by their colleagues. On the other hand, if the principles and ethical obligations rely on such a mechanism, there should be concern as to whether it can be effective. Dissemination of the codes is more problematic. Each organization requires its members to acknowledge and to abide by its code of ethics, but the amount of ethics training or education varies significantly. ASCE has a downloadable manual available on its website and includes links and resources for members with questions, as well as links to other organizations and nonengineering ethics resources. My phone inquiries were handled without hesitation, and I was directed to the correct knowledgeable staff person. ASLA, AIA, and ASCE all include an ethics problem in their monthly communications to members, and each of them frame the solution to the problem in the relevant sections of its particular code; however, each relies on an interested reader for dissemination. ASCE has formed a Foundation of Professional Practice that includes an active ethics awareness program for participating members and firms and has an "Ethics Hotline," which members can use to discuss ethics concerns that they may have about an action or decision. None of the peak organizations considered has an active program of required ethics training, and it is unlikely that the organizations could compel members to participate in such a program. ASLA, AIA, and ASCE offer ethics programs from time to time in various formats.[25]

Sustainability and the Ethical Challenges for Designers

Society sanctions the profession because of the perceived benefit of the profession's activities and the desirability of the outcomes of those activities. In exchange for the sanction, society expects a certain level of performance from practitioners, a degree of what is generally described as "professionalism." Ladd observes that "...members of a profession are considered to have special duties and responsibilities towards society...."[26] It might be said that it is simply by virtue of the special knowledge and skills that an individual would have these special duties. In fact, every individual has a right and an obligation to pursue her own well-being to the extent that such acts do not harm another. Further, every person has an obligation to make some sacrifice to prevent harm to another person or to assist a person in need. People that have developed special skills and are afforded social advantage because of them acquire a greater responsibility than those who have not.[27] Designers have special training and knowledge that allow them to better anticipate and assess the effects of actions than nonprofessionals.

Some suggest that designers have no such obligation. For example, Florman argues that engineering as a profession no longer controls itself as a profession from within. Laws and regulations have encoded into statute what was once the purview of professional ethics and group enforced behavior. Moreover, most engineers are employed by large organizations and do not operate as autonomous individuals in the marketplace. Let us note that though Florman wrote specifically of engineers, his arguments and the arguments that follow from them could be directed to any of the design professions. In his view, the engineer is in no position to judge what projects are good or bad, right or wrong. These are issues to be resolved by society at large, by elected officials, by the courts. In his view, engineers should not have the right to choose for whom they work; in essence. engineering in itself is neither moral nor immoral. "It is not the engineer's job, *in his or her daily work*, (italics in the original) to second guess prevailing standards of safety or pollution control, not to challenge democratically established public policy."[28] Florman argues that by performing their tasks competently, designers are being virtuous. He argues "If we relate engineering ethics to protection of the public interest, then clearly diligence is more moral than conventional 'morality'."[29]

Florman wants it both ways. He wants to excuse engineers from making the hard calls and looking beyond current law and public policy, but then, he wants us to find the competent engineer that merely follows the law as well as is virtuous and "more moral" than his fellow citizens. If they are to be worthy of the sanction of the community, the professionals must be subject to a duty or obligations greater than other citizens. By virtue of the special knowledge of their profession, they acquire the burden to make ethical judgments and to act in the public's interests.

In response to Florman, Alpern writes that "though engineers are bound by no special moral obligations, ordinary moral principles as they apply in the engineer's circumstances stipulate that they nonetheless be ready to make greater personal sacrifices than can normally be demanded of other individuals."[30] He makes his argument first by describing what he calls the Principle of Care—"Other things being equal, one should exercise due care to avoid contributing to significantly harming others."[31] This includes knowing what harm might result from our actions and taking steps to mitigate or avoid that harm. It refers to one "contributing" to harm, so the obligation extends to taking part in activities that might result in harm.

Alpern acknowledges that the Principle of Care is vague; the degree of care required is not always the same. He writes "the degree of care due is a function of the magnitudes of the harm threatened and the centrality of one's role in the production of that harm."[32] He introduces the Corollary of Proportionate Care: "When one is in a position to contribute to greater harm or when one is in a position to play a more critical part in producing harm than is another person, one must exercise greater care to avoid doing so."[33] In light of the Principle of Care and the Corollary of Proportionate Care, and of the role of professionals, they should be held to a higher standard of care; that is, they must be willing to make greater sacrifices than others for the sake of public welfare. Professions are valuable to society because of their commitment to the public good, and the professions acknowledge this. The code of ethics in every design professional organization includes in its canons that the professional "shall hold paramount the safety, health and welfare of the public..."[34] or words to that precise effect.

It is expected then that the standard of care for professions will change to reflect the society in which it is operating. "All professionals are moral enterprises that involve concerns beyond the application of technical principles. How well the professionals meet these moral obligations will determine the freedom of the individual professional enterprise."[35] Indeed, as social concerns with the environment have increased, the professions have responded to public criticism of their performance by both digging their heels in and resisting change in some quarters and with inspiring leadership in others.

A limited survey of the canons of practice and codes of ethics completed for this book of selected professional organizations representing design or planning professionals found a reference to the commitment to public interest in every case. In some cases, the canons of professional ethics included a statement about the environment. The character of these statements varied substantially from "shall consider the environmental impact"[36] and "strive to protect the integrity of the natural environment"[37] to the more robust "strive to comply with the principles of sustainable development in the performance of their professional duties."[38] Some organizations have statements regarding the standard of care regarding environmental impacts and go further by specifically including negative obligations for the professional to prevent harm. ASCE, for example, includes such a statement. The character of these statements was considered so onerous by some members of ASCE that they withdrew from membership when the statement was first added.[39] ASLA also includes statements that together create negative obligations requiring the professional to prevent harm. The current ASCE code of ethics has a robust statement regarding environmental protection or quality and sustainability. Canon 1 of the code of ethics states that engineers "shall strive to comply with the principles of sustainable development in the performance of their professional duties."[40] It goes on to provide a definition of sustainable development "Sustainable Development is the challenge of meeting human needs for natural resources, industrial products, energy, food, transportation, shelter, and effective waste management while conserving and protecting environmental quality and the natural resource base essential for future development."[41]

The ASCE code provides an illustration of an important element to be remembered; codes change as the professions change. The original ASCE code passed in 1914 was fairly straightforward and consisted of six different areas of behavior prohibited for members. In general, it described how members of the society were to act toward each other and their respective clients. There was no mention of an obligation to protect public health and safety or concern over the welfare of the public. The 1961 version of the ASCE code of ethics was longer but, in essence, was a more detailed list of proscribed acts. The duty to the public's interests was still not mentioned. By 1977, the format of the code of ethics changed to include four fundamental principles and seven fundamental canons. The first fundamental canon reads "Engineers shall hold paramount the safety, health and welfare of the public in the performance of their professional duties."

By 1993, under the Guidelines to Practice Under the Fundamental Canons, item (f) reads "Engineers should be committed to improving the environment to enhance the quality of life,"[42] essentially linking environmental quality and public health and safety within the code of ethics. In 1996, ASCE revised the fundamental canon to read "Engineers shall hold paramount the safety, health and welfare of the public and shall strive to comply with the principles of sustainable development in the performance of their professional duties." This change was followed in 1996, with the definition of sustainable development (quoted above), included as an endnote. Another revision occurred in 2006, where an endnote was added to the title "Code of Ethics," which states that "The Society's Code of Ethics was adopted on September 2,1914 and was most recently amended on July 23,2006. Pursuant to the Society's Bylaws, it is the duty of every Society member to report promptly to the Committee on Professional Conduct any observed violation of the Code of Ethics."[43] These modifications, and the most recent version in 2017, to the ASCE code of ethics illustrate the dynamic nature of the code and the efforts taken by the peak organization to maintain the profession's ethical standards.

In 2008, ASLA enacted a code of environmental ethics to guide the professional behavior and activities of their members. This is a lengthy and substantive description that declares that ASLA members should approach their work, clients, and stakeholders with a clear intent to foster sustainable outcomes. (See Appendix A for the entire text.) It is important to note that the language of the code of environmental ethics uses the world "should" rather than "shall," which is used in the ASLA code of ethics. This same sort of arrangement is present in the AIA code, where rules are enforceable but canons are more akin to recommendations. This reduces the enforceability of the environmental code but not its significance. These steps by ASCE, AIA, and ASLA reflect the process of gradual change and adaptation to changes in the professions that reflect the growing awareness of environmental distress and population pressures that are before us. These are examples of powerful and substantive changes, because they are a consensus from within the professions. It should be noted that except for the Industrial Design Society of America, all of the organizations shown in Table 2.2 have updated or revised their respective codes of ethics since the first edition of this book. No evidence of revisions or update was found for the Industrial Design Society of America.

Table 2.3 summarizes the aspects of selected peak associations of various design professions. It must be noted that every code of ethics included a commitment to safeguard the health, safety, and welfare of the public. This concern is the threshold commitment of what it means to be a professional.

TABLE 2.3
Environment and Sustainability Considerations in Codes of Ethics of Selected Peak Organizations (Revised 2018)

Peak Association	Year	Public Health, Safety, and Welfare	Protection of the Environment	Duty to Report Violations or to Prevent Harm	Specific Sustainability Statement
American Association of Engineering Societies[a]	2017	Yes	Yes	No	Yes
American Consulting Engineers Council[a]	Unknown	Yes	No	No	No
American Institute of Architects	2017	Yes	Yes	Yes	Yes
American Institute of Chemical Engineers	2015	Yes	Yes	Yes	No
American Planning Association/American Institute of Certified Planners	2005	Yes	Yes	No	No
American Society of Civil Engineers	2017	Yes	Yes	Yes	No
American Society of Landscape Architects	2015	Yes	Yes	Yes	Yes
Industrial Design Society of America	1998	Yes	Yes	No	No

(Continued)

TABLE 2.3 (Continued)
Environment and Sustainability Considerations in Codes of Ethics of Selected Peak Organizations (Revised 2018)

Peak Association	Year	Public Health, Safety, and Welfare	Protection of the Environment	Duty to Report Violations or to Prevent Harm	Specific Sustainability Statement
Institute of Electrical and Electronic Engineers	Unknown	Yes	Yes	Yes	No
National Association of Realtors	2014	Yes	No	No	No
National Society of Professional Engineers	2007	Yes	Yes	Yes	Yes
Society of Manufacturing Engineers	Unknown	Yes	Yes	No	No
American Society of Mechanical Engineers	1998	Yes	"…shall consider environmental impact…"	Yes	No
Chinese Association for Science and Technology[a]	1996	Yes	Yes	No	No

(Continued)

TABLE 2.3 (Continued)

Environment and Sustainability Considerations in Codes of Ethics of Selected Peak Organizations (Revised 2018)

Peak Association	Year	Public Health, Safety, and Welfare	Protection of the Environment	Duty to Report Violations or to Prevent Harm	Specific Sustainability Statement
Chinese Association for Science and Technology	2014				
Academy of Design Professionals	2017	"...should promote and serve the public interest..."	"...shall consider environmental, economic, and cultural implications..."	No	No
European Council of Civil Engineers	2000	Yes	Yes	No	Yes

[a] Chinese Association for Science and Technology, American Association of Engineering Societies, and American Consulting Engineers Council are organizations of organizations or firms and do not directly represent individual practitioners; therefore, their respective codes or standards are not focused directly on the behavior of individuals.

The most comprehensive concern with the environment in general and sustainability specifically is found in the canons of AIA, ASCE, and ASLA. All of these canons include a duty to report actions that may cause harm; of the others, only the American Institute of Chemical Engineers and the Institute of Electrical and Electronic Engineers (IEEE) included a statement akin to this. AIA and ASLA have robust statements within their professional codes of conduct describing far-reaching obligations, ranging from design to professional practice. The Industrial Designers Society of America (IDSA) has a separate, fairly robust environmental policy statement that encourages its members to be aware of and to employ environmental concerns in their work. In contrast, the American Society of Mechanical Engineers (ASME) is virtually silent regarding the environment or sustainability in its code of ethics.

The ASCE guidelines to practice include guidance for engineers "to seek opportunities to be of constructive service in civic affairs and work for the advance of the safety, health and well-being of their community" and to have a commitment "for improving the environment to enhance the quality of life."[44]

The language of the various codes of ethics is interesting to think about. Several of these organizations have membership made up primarily of professional practices or businesses, as opposed to individual practitioners. The nature of these organizations is to promote the business of the professions, and while they do have published codes of ethics, the focus tends to be on the behavior of the business organization and less so on the nature of individual design decisions or outcomes. Perhaps for this reason, statements regarding the environment and sustainability are absent for the most part. In other cases, the language is crafted carefully. For example, the National Society of Professional Engineers (NSPE) includes a professional obligation in its rules of practice that "Engineers shall at all times strive to serve the public interest" and that "Engineers are encouraged to adhere to the principles of sustainable development in order to protect the environment for future generations." The latter statement includes a footnote to direct the reader to the following statement: "'Sustainable development' (quotes in the original) is the challenge of meeting human needs for natural resources, industrial products, energy, food, transportation, shelter and effective waste management while conserving and protecting environmental quality and the natural resource base essential for future development." It is unclear where the professional obligation lies in this statement, since it only encourages the professional to strive and adhere to the somewhat vague standard outlined in the footnote. Like other definitions of sustainability, it is difficult to sort out the meaning when it comes to practice, though NSPE should be given credit for trying to define it. It also points out the challenge to the individual designer; what does this actually mean in day-to-day practice?

In this second edition of the book, some effort was made to include design organizations from different countries, primarily in the tables of this chapter. A direct comparison of the content of codes of ethics of design professions in Western economies and traditions is difficult, since Asian professions seem to organize around "Guiding Principles" that do not align directly with Western codes of ethics. While both standards articulate the expectations and values of appropriate and ethical professional performance, they are rooted in very different traditions. In general, Western ethical tradition is centered in the duties and actions of an individual and are rooted in values. In a sense, their ethical obligations have a closely held personal sense of obligation and duty. Asian cultures tend to have a collective view of ethics—"good" is determined by what is best for the group, as opposed to the individual.[45] Research has shown that managers in Western-oriented firms tend to value their personal ethics more than the ethics of their supervisors or the organization. Asian managers may see their duty to support their manager or leadership rather than to their own values.

This complicates a direct comparison. In a sense, both are value-based standards of care, but they reflect different cultural, social, and governmental experiences. A more complete comparison is beyond the scope of this effort, but as China rises as an economic force and engages in ever-greater efforts in global development, some effort to understand the differences should be undertaken. A review of some of the professional design organizations in the People's Republic of China shows that they share "assessment standards,"[46] which appear to be akin to performance standards such as those published by the American Standards for Testing and Materials (ASTM), as opposed to ethical guidelines found in similar Western organizations. See Appendix B for an expanded consideration of this.

The variance in the character of the statements and absence of such statements from some professional codes of ethics may merely reflect that we are in a period of change and some groups have adapted more quickly than others. Alternatively, it may be that the hesitation to embrace this responsibility for environmental issues is a reflection of business concerns dictating professional conduct. Given the character of these statements, especially where there is a negative obligation to prevent harm, we should expect to see clear differences in the work products as the design professions come to terms with the values that support sustainable design. If there is failure to act on these aggressive standards of professional behavior, it may become apparent to the marketplace that the professions are not capable of policing themselves. In such a case, the likelihood that laws will be passed or public policy emerge to address the lack of leadership and to mandate a threshold for an appropriate standard of care. Such legislation, when it occurs, might be seen as the public, through its institutions, expressing its interests or as a rebuke to the design professions for a lack of leadership. In either case, the laws codify the expected minimum level of performance, in effect a standard to be met rather than exceeded. This threshold often becomes the measure of

Professional Ethics 43

performance and constrains creativity and, in effect, defines acceptable standard of care. This in turn replaces ethical behavior with design by the standard or, as is sometimes described, with "cookbook" design.

A reading of the literature of the ethical obligations of design professionals provides some insight into how the individual designer might approach the question of obligations to the environment and sustainability. Both Spier[47] and Florman[48] were practicing professionals at the time they were writing, and they assert that the design professional cannot be held accountable for his work as morally good or bad. Spier argues that the dual nature of technology is such that the designer cannot be held responsible[49]; that is, all technology, and therefore design, has both positive and negative aspects and that design is a two-edged sword. The hammer that is used to build the house can also serve as an effective weapon. The design of a weapon of self-defense is just as lethal as an offensive weapon, when it is used. This sort of dilemma is addressed in the rules of conduct for members of the British organization Institute of Engineering and Technology (IET). Members are directed to "take all reasonable steps to avoid the waste of natural resources, damage to the environment and damage or destruction of man-made products," but an exception for actions taken in the national defense.[50] It seems reasonable that some similar allowance is to be made when the protection of the public welfare is at hand.

On the other hand, Florman believes that ethical standards for designers have been replaced by law and consensus standards and that the designer has no burden beyond compliance with public policy, technical standards, and professional competence. Both suggest that designers cannot fairly deny services to anyone simply because they disagree with the project; in essence, people have a right to design services.[51] More so, they argue, when the engineer is not acting as a sole practitioner but as an employee of a large organization, the individual's professional responsibility is subject to the scope of his individual work and not to the activities of the firm as a whole.

Ladd agrees that a professional has no more knowledge of ethical behavior than a client, at least by virtue of the profession alone. While there are no special ethics for professions, he asserts that they do have greater obligations that stem from the power of being a professional; both individual and collective powers of professions beget responsibility.[52] The consensus of professional organizations differs with Spier and Florman. In support of the focus of the United Nations (UN) on sustainable development, the World Federation of Engineering Organizations prepared a summary of the actions taken or planned by various engineering organizations around the world in support of or in response to the UN conference in Rio de Janeiro. Of particular interest were the recommendations for change in future engineering activities.[53] The report recommends that the following steps be included in all future engineering activities:

- Environmental studies be performed as part of all relevant projects. Such studies will normally require a multi-disciplinary approach.

- Evaluate the positive and negative environmental impacts of each project. This evaluation might be based on a preliminary view of the available information or the engineer's experience. They should evaluate the basic function and purposes behind a project. They should suggest alternatives to clients if environmental risks emerge.
- Develop improved approaches to environmental studies. Environmental effects should be considered early in the planning process. Studies should evaluate the long-term consequences of environmental changes.
- Make clients aware that engineers can reduce but not always eliminate adverse environmental impacts. The legal and financial responsibilities of all parties should be clearly defined.
- Urge clients to prevent or minimize the diverse environmental effects of projects in all phases—initial planning, design, construction, commissioning, operating, and decommissioning.

Bibliography

American Consulting Engineers Council Code of Ethics, http://www.iit.edu/departments/csep/PublicWWW/codes/coe/acec-b.htm

American Consulting Engineers Council Professional and Ethical Conduct Guidelines, http://www.ooonlineethics.org/codes/acec.html

American Institute of Architects Code of ethics and Professional Conduct, http://www.iit.edu/departments/csep/PublicWWW/codes/coe/American%20Institute%20of%2Architects.htm

American Society of Civil Engineers Code of Ethics, 1996, http://wwwasce.org/about/codeofethics.cfm

American Society of Landscape Architects ASLA Code of Professional Ethics, 2001, http://www.asla.org/nonmembers/CODE499.htm

American Society of Landscape Architects, ASLA Declaration on Environment and Development, 1993, http:www.asla.org/nonmembers/declarn_env_dev.html

Beauchamp, Tom L., and Norman E. Bowie, *Ethical Theory and Business*, 5th edition, New York: Prentice Hall, 1997.

Maitland, Ian, "The limits of business self-Regulation." in Beauchamp, Tom L., and Norman E. Bowie, *Ethical Theory and Business*, 5th edition, New York: Prentice Hall, 1997.

Rafalko, Robert J., "Remaking the corporation: The 1991 U.S. sentencing guidelines." in Beauchamp, Tom L., and Norman E. Bowie, *Ethical Theory and Business*, 5th edition, New York: Prentice Hall, 1997.

Endnotes

1. Singer, Peter, *Practical Ethics*, 2nd edition, Cambridge University Press, Cambridge, UK, 1999, p. 234.
   Singer says we have an obligation to assist another (even when survival is at issue) he *ought* to assist, that is he is ethically required to assist. Singer's point in essence recognizes our dependency on one another, we are not "isolated individuals" (p. 629). His argument to assist is based on this interdependency ...we recognize that if we can prevent a bad thing from happening at some relatively small costs then we ought to prevent it. Once a thing is seen as bed then we are obligated to do what we can to prevent it or some degree or portion of it.

2. Donaldson, Thomas, The ethics of conditionality in international debt, *Ethical Theory and Business*, Tom L. Beauchamp and Norman E. Bowie, editor, Prentice Hall, Upper Saddle River, NJ, 1997, p. 646. Donaldson writes of correlative duties, that is we are required to do more than simply the obligation to not violate the rights of another—the existence of rights and the force of moral judgment require us to (1) avoiding depriving; (2) help protect others from deprivation; and (3) aid the deprived. Donaldson says businesses (in his argument multinational businesses) have obligations for the first two duties while governments have obligations to all three.

3. The text references the American Medical Association's role as essentially a trade union of sorts. In the United States licensure is done by the states, with a few exceptions. Although the AMA is a powerful organization and does act at least in principle for the benefit of its members it does not in fact license physicians. This arrangement is true of all of the professions with the exception of lawyers and accountants oddly enough. To be sure the AMA plays an important role on licensure boards but most laws require a balance mix of licensure board members including a fairly broad range of stakeholders not only those who carry the license. This is an important distinction to made in light of the text's concern with the AMA as an example, but merely begs the larger question at hand.

4. ibid. From Discussion Questions on page 171.

5. "Attributes of a profession." Ernest Greenwood in *Ethical Issues in Engineering*, Johnson, Deborah, editor, Prentice Hall, Englewood Cliffs, NJ, 1991, pp. 67–77.

6. Firmage, D. Allen, "The definition of a profession." in *Ethical Issues in Engineering*, Johnson, Deborah, editor, Prentice Hall, Englewood Cliffs, NJ, 1991, p. 63.

7. Ladd, John, "Collective and individual moral responsibility in engineering: Some questions." in *Ethical Issues in Engineering*, Johnson, Deborah, editor, Prentice Hall, Englewood Cliffs, NJ, 1991, pp. 26–39.

8. Licensed design professionals are given special advantage in the marketplace by virtue of their license. The state or licensing authority limits who can perform these tasks in the marketplace and for the most part it gives the control over who is licensed to the professions themselves. In this way the professions have influence, if not control, over who and how many join their ranks. Once among the community of sanctioned professions practitioners are held

to account primarily by others in their profession. Courts differ to the special knowledge of professionals when determining what is an appropriate standard of care, what is negligence etc.

[9] Ferree, George W. and Scott R. Vaughan, *The Moral Obligation of the Gifted*, Michigan State University, East Lansing, MI, www.msu.edu/~vaughan/mog.html

[10] Beauchamp, Tom L. and Norman E. Bowie, *Ethical Theory and Business*, 5th edition, Prentice Hall, New York, 1997, pp. 117–118.

[11] I am referring to the breakdown in corporate and professional ethics in the various business failures in the United States, most notably Enron and WorldCom and the more recent banking failures. These failures are attributed to unethical, sometimes illegal business practices. In both cases professionals from within the corporation and hired as consultants failed to comply with widely accepted business practices, as well as professional codes of ethics regarding conflicts of interest and disclosure.

[12] Maitland, Ian, "The limits of business self-Regulation" in Beauchamp, Tom L. and Norman E. Bowie, *Ethical Theory and Business*, 5th edition, Prentice Hall, New York, 1997, p. 127.

[13] ibid., p. 132.

[14] ibid., p. 134.

[15] Rafalko, Robert J., "Remaking the corporation: The 1991 U.S. sentencing guidelines" in Beauchamp, Tom L., and Norman E. Bowie, editor, *Ethical Theory and Business*, 5th edition, Prentice Hall, New York, 1997, p. 156.

[16] ibid., pp. 160–161.

[17] These seven steps are discussed at length in Rafalko's essay. Such discussion is illustrative but beyond the scope of this book.

[18] Layton Jr., Edwin T., *The Revolt of the Engineers, Social Responsibility and the American Engineering Profession*, Johns Hopkins Press, Baltimore, MD, 1986. Layton lays the groundwork for an argument similar to Adam Smith's invisible hand. If the engineer is engaged in his profession and acting as a good citizen to achieve his own goals, society will be improved incidentally by the good works of the engineer. p. 62.

[19] In my own experience questions of ethics did occur in the course of professional practice and were usually addressed by appealing to the senior member of the firm present. In most cases the situation was resolved without an appeal to either the code of ethics or moral reasoning but rather to past practice and the experience of the senior person.

[20] Although the design professional may be construed as included other groups of professionals, I have arbitrarily selected a handful of peak organizations for consideration. I have selected these on the basis that primary groups of professions are represented (architects, engineers, landscape architects, industrial designers) and a business organization of professional firms.

[21] By way of disclosure, I am a licensed landscape architect, a former Member of ASLA and have in the past served as company representative to and a board member of the Maryland Chapter of ACEC.

[22] ASLA, IDSA and ASCE have fairly lengthy statements regarding the environment and sustainability contained in the Codes of Ethics or included by reference.

[23] ASCE Fundamental Canons 4. States Engineers shall act in professional matters for each employer or client as faithful agents or trustees, and shall avoid conflicts of interest. The Fundamental Canons and the Guidelines to Practice are silent as to the limits of this statement. American Society of Civil Engineers Code of Ethics, 1996, http://wwwasce.org/about/codeofethics.cfm

[24] American Society of Landscape Architects ASLA Code of Professional Ethics, Canon 1 Professional Responsibility Rule 1.109, http://www.asla.org/nonmembers/CODE499.htm

[25] This information gathered through a search of websites and a familiarity with these organizations.

[26] Ladd, John, "Collective and individual moral responsibility in engineering: Some questions." in *Ethical Issues in Engineering*, Johnson, Deborah, editor, Prentice Hall, Englewood Cliffs, NJ, 1991, pp. 26–39.

[27] Ferree, George W. and Scott R., *Vaughan The Moral Obligation of the Gifted*, Michigan State University, East Lansing, MI, www.msu.edu/~vaughan/mog.html

[28] Florman, Samuel, *The Civilized Engineer*, St. Martin's Press, New York, 1987, p. 99.

[29] ibid., p. 102.

[30] Alpern, Kenneth D., "Moral responsibility for engineers." in *Ethical Issues in Engineering*, Johnson, Deborah, editor, Prentice Hall, Englewood Cliffs, NJ, 1991, p. 187.

[31] ibid., p. 188.

[32] ibid., p. 189.

[33] ibid., p. 189.

[34] American Society of Civil Engineers, *ASCE Code of Ethics*, American Society of Civil Engineers, effective November 1996, www.asce.org/sbout/codeofethics.cfm 3/2/02

[35] Firmage, D. Allen, "The definition of a profession." in *Ethical Issues in Engineering*, Johnson, Deborah, editor, Prentice Hall, Englewood Cliffs, NJ, 1991, p. 65.

[36] *Code of Ethics of Engineers*, American Society of Mechanical Engineers, 3/2/02, www.onlineethics.org/codes/ASMEcode.html

[37] *Ethical Principles in Planning*, American Planning Association, effective May 1992, 3/2/02, www.iit.edu/de[partments/csep/PublicWWW/codes/coe/

[38] *ASCE Code of Ethics*, American Society of Civil Engineers, effective November 1996, www.asce.org/sbout/codeofethics.cfm 3/2/02

[39] Conversation with Thomas Smith, ASCE Council, June 2002.

[40] ASCE Code of Ethics, http://www.asce.org/inside/codeofethics.cfm, June 2008

[41] ASCE, ibid.

[42] ASCE, http://ethics.iit.edu/codes/coe/amer.soc.civil.engineers.1993.html, Institute for the Study of Professions, June 2009.

[43] ASCE, Code of Ethics, https://www.asce.org/inside/codeofethics.cfm, June 2009.

[44] ASCE Code of Ethics, American Society of Civil Engineers, effective November 1996.

[45] Fredi Garcia, Diana Mendez, Chris Ellis, Casey Gautney, "Cross-cultural, values and ethics differences and similarities between the US and Asian countries," *Journal of Technology Management in China*, 9(3), 303–322, 2014, https://doi.org/10.1108/JTMC-05-2014-0025 https://www.emeraldinsight.com/doi/full/10.1108/JTMC-05-2014-0025 accessed June 1, 2018.

[46] Xu Chunmei, personal email, May 22, 2018.
[47] Spier, Raymond, *Ethics, Tools and the Engineer*, CRC Press, New York, 2001.
[48] Florman, ibid.
[49] Spier's perspective is that issues faced by engineers are complex and for the most are to be resolved by society not engineers. His key point is that a given technology can be used for good or bad and it is this duality that excuses the engineer from moral responsibility in Spier's opinion. Much like Florman, he believes the engineer cannot be expected to decide what is good and what is not; this responsibility lies somewhere else.
[50] Institute of Engineering and Technology, Rules of Conduct No. 8 , www.theiet.org
[51] This argument is without merit. If the designer is not free to exercise choice than how can he be free to exercise the judgment necessary for professional standard of care.
[52] Ladd, John, "The quest for a code of professional ethics: An intellectual and moral conflict." in *Ethical Issues in Engineering*, Johnson, Deborah, editor, Prentice Hall, Englewood Cliffs, NJ, 1991, pp. 130–136.
[53] The Engineer's Response to Sustainable Development, The World Federation of Engineering Organizations, January 10, 1997, http://www.ecoucil.ac/rio/focus/report/english/wfeo.htm

3

Is There an Ethical Obligation to Act Sustainably?

The most erroneous stories are those we think we know best—and therefore never scrutinize or question.

Stephen Jay Gould

Theories of Ethics

To provide a basis for the arguments presented in this book, it seems there should be an understanding of the theories of ethics in general, before engaging in a deeper discussion of professional ethics. It is presumed that the readers have a general knowledge of ethics as they relate to civil behavior and perhaps a working knowledge of business or professional ethics. A general description of the more common theories of ethics is provided here to both rekindle the academic appreciation of ethics and provide a glimpse into how these ideas may be used in this book. In a sense, this overview describes the language of ethics and highlights important concepts addressed in the following pages.

Up to now, ethics have been humanist in purpose and character. We have reached a point where our casual impact and activities have profound effects on the environment and where the gap between the first and third worlds is growing. Thomas Berry correctly observed that the argument is usually framed as the good of nature against the bad of society or as the bad of society against the good of society. Berry suggests each side has good and bad points, and we should compare the good with the good and the bad with the bad.[1] We need to answer the question: should ethics extend the cultural limits further to include eccentric elements—respect for life diversity and community?[2]

Normative theories of ethics are about principles for distinguishing right from wrong. In general, normative ethical theories can be distinguished as either consequentialist or nonconsequentialist. Consequentialists believe that the moral rightness of an action is determined by its outcome: if the outcome is good, then the act is right and moral. Nonconsequentialist (deontological) theories contend that right and wrong are determined by more than the consequences of an action. Consequences may be important, but nonconsequentalists believe that other factors are also relevant to moral assessment. Professional ethics tend to be consequentalist in nature. In other words, it is the outcome of the design that is weighed, not the underlying rules or motivations for the design.

Questions about consequentalist ethics usually deal with concerns about who is the outcome good for? Should consequences be considered for all or just for the actor? This is particularly important when we consider the negative obligations included in some professional codes of ethics that require the professional to prevent harm. Who in the array of stakeholders in a given design is in a position to define harm? Of course, designers working in a marketplace must provide positive outcomes beyond their own, or they will not be working for very long; the work of designers has multiple stakeholders. The stakeholders include the client, those who build or manufacture the product of design, those who distribute and sell the product, and the end user. The stakeholder shortlist is frequently expanded to include the marketplace, the designer's own profession, and the interests of the public at large. So, design is evaluated by how well the designer's work meets the expectation and the needs of the various stakeholders. How well the designer satisfies this need determines the design to be good and an ethical act; to the degree it fails to satisfy, it is not good and not ethical from a professional sense. So, from a consequentialist view, the end product is an ethical statement of sorts.

Moral rules are fundamentally derived from the obligation to respect every human, but moral obligations may be derived from many sources. Consider, for example, WD Ross's prima facie duties: a prima facie obligation is an obligation unless it is overridden by more pressing obligations.[3] McMahon goes further to assert an argument for moral intuition, a "spontaneous moral judgment."[4] Such intuition enables a person to make a moral judgment in circumstances not previously encountered. Intuition, on the other hand, is likely to be guided by our learned moral framework rather than be a wholly new moral paradigm. Moral rules are not static. Our judgment reflects our experience and our interests. As conditions change and as we learn as individuals and then as cultures, our moral standards change to reflect our new circumstance. In the past slavery, institutionalized class systems and cruelty to animals were accepted as moral, but over time, the moral barometer changed, as the interests of the society changed. This expanding moral umbrella is a function of the security and stability of the human species and cultures. The scope of inclusion expands to cover more people and living things as our fates intermingle. Our moral vision includes others, at least in part, as a function of our ability to meet our own needs first (Maslow's Hierarchy of Needs). In the absence of those needs, our social scope of moral inclusivity shrinks. Likewise, as our needs are met and secured, the aperture of moral standards opens.

Morality, it seems, changes then to reflect our awareness of an objective thing or circumstance and an understanding of how we relate to it. Our concern is lately expanding to include our environment.[5] If, however, morality is based on a cogent evaluation of our behavior, it is reasonable to assume that some knowledge, appreciation, or awareness of the environment is required.

In general, ethics can be evaluated in three ways: in terms of the Kantian Categorical Imperative, in terms of virtue morality, and in terms of utility. Emmanuel Kant (1724–1804) posited that we should always act in such a way that we can will the maxim (a principle of action) of our action into a universal law. A thing is morally right if and only if we can "will it to become a universal law of conduct." A universal law in Kant's view is understandable and can be discovered through reason by anyone, and it would apply to all people all of the time. Kant believed that the views of morality held by most people are pretty much correct and that all rational beings will rightfully pursue their own happiness. According to Kant, when people act for a reason, that is, they decide to act, they are following a maxim, a rule of action. He did not believe that each of us establishes a rule of action for each decision but that we can discover our reasons for acting if we were to stop and think about it. A maxim takes the form of "I will do action X in circumstance C for purposes P." Kant tested maxims by looking for contradictions in their universality. Inconsistency can arise in a maxim as a universal law in two ways: (1) if, as a universal law, it would jeopardize our survival or happiness, and (2) if we will something to be a universal law that cannot be a universal law. Kant believed that we have a duty to do the moral thing and that the decision to do the moral thing must be based on the duty, not on a presupposition of a favorable outcome or some other motivation; doing the right thing is the right thing to do. He crafted the Categorical Imperative to express this: "Act so that you treat humanity, whether in your own person or in that of another, always as an end and never as a means only."

The Categorical Imperative is an unconditional directive: do this, or don't do that. What makes an action right is that the agent would be willing to be so treated if the positions of the parties were reversed and that human beings are treated as ends in themselves rather than means—in short, the familiar "golden rule." Several duties are derived from the Categorical Imperative: everyone has a duty to help those in need, everyone has an obligation to express gratitude because of a benefit one receives, and everyone has a general respect for other people.

A more recent distillation of these ideas is found in the work of John Rawls (1921–2002), who was concerned about justice as it is expressed in our actions. John Rawls wrote that the correct standards of justice are those that rational, self-interested individuals in a certain hypothetical state analogous to the state of nature (the original position) would accept if given the choice. Key to Rawls's theory is that the individual does not have to accept the standards; it is enough only that he would accept them if the choice existed. We can extend Rawls's ideas to moral and ethical choices. In this view, it is important that the only acceptable moral rules or ethical acts are those that everyone can agree on and that can be publicly justified. The hypothetical process that Rawls proposes makes his approach interesting; he requires that there be a "veil of ignorance" drawn over the members of society; that is, as they describe these rules of moral and ethical behavior, they can have no personal information of themselves. All decisions about principles are made behind a veil of ignorance, which assures impartiality and objectivity. Rawls says that all rational, self-interested individuals in the original position acting behind a veil of ignorance would agree on two principles to regulate the distribution of freedoms and economic resources in society:

1. Each person is to have an equal right to the most extensive basic liberty compatible with a similar liberty for others.
2. Social and economic inequalities are to be arranged, so that they are both (a) the greatest benefit of the least advantaged and (b) attached to offices and positions open to all under conditions of fair equality of opportunity.

Moral rules and ethics are derived from these obligations to respect every human.

Virtue ethics are derived from Greek philosophy, wherein the primary function of morality is to cultivate virtuous traits in a person and virtues are acquired through practice. Like Kant would believe, the acquisition of virtues and moral behavior depend upon appropriate motivation. If one is raised to be virtuous, one will act in ways consistent with those virtues. Aristotle believed that it was the obligation of the state to educate its citizens in the various virtues because it had an overarching interest in having virtuous citizens. A virtuous person acts on the learned virtues as his or her own reward. A virtuous person acts in an ethical manner, because to do otherwise would be inconsistent with the virtues he has embraced and learned.

Utilitarian ethics are concerned with producing the greatest possible good or happiness for everyone. Since utilitarians evaluate moral goodness based on the consequences, almost anything may be morally right or wrong, depending on the circumstances. Utilitarians also wish to maximize happiness over the long run, so that some early reduction in happiness may be acceptable if there is the promise of greater future happiness. Since we cannot know the future with certainty, we must act on what is likely to produce the best outcome. Although we are concerned with the total happiness, we do not have to ignore our own happiness in the process.

Utilitarianism is concerned with outcomes, but when considering the greatest happiness, we must also consider unhappiness. Utilitarians are often categorized as either act or rule utilitarians. Act utilitarianism considers an act right if and only if the good or benefit of the act outweighs the bad of the act. Also, good and bad are determined not solely in the terms of the actor but for the benefit or harm of all. It requires the moral agent to think of the consequences for all. Rule utilitarianism, on the other hand, distinguishes between an act and a rule. Rule utilitarianism requires that moral rules be observed and that these rules be followed based on the greatest utility or tendency to promote happiness.

Jeremy Bentham (1748–1832) used the concept of utility to describe the tendency of an act or a thing to increase or decrease happiness. Morality requires an actor to pursue the common good rather than his or her own happiness exclusively, but the notion of a common good is vague. Good and bad differ from person to person. Bentham tried to refine the notion of a common good. He wrote that the community does not exist as a body of members, but rather, the community exists as the "sum of the interests" of the members. In his view, then, a thing is good if it advances or promotes the interests and bad if it inhibits or constrains the interests. Among possible actions, any action that increases the interests/happiness of the whole would be a moral choice.

John Stuart Mill (1806–1873) argued that rightness and wrongness are relative and occur as matter of degree: an act is right in proportion to its tendency to promote happiness. The more happiness it will produce, the more right it is. This is known as Mill's Greatest Happiness Rule. The standard, though, is not the agent's own happiness, but the greatest amount of happiness taken altogether. Mill suggested that we should follow moral rules and consider the utility of a choice only when there is a conflict between rules.

Problems arise if act utilitarianism might require a person to forego his or her own interests or happiness in favor of increasing the total happiness. Bentham suggests a negative formulation of the Greatest Happiness Principle, requiring instead that an action is wrong if and only if it would reduce total happiness and there are alternatives that would not reduce total happiness. Act utilitarianism ignores the distribution of happiness; as long as the greatest utility or happiness is achieved, the distribution is important.

Critics of utilitarianism question whether it is workable. Is it possible to even know enough to be able to determine the greatest happiness or unhappiness? Further, can something be wrong even if it produces the greatest good? Utilitarians might argue that the greatest happiness for most people is moral and ethical, regardless of great poverty or deprivation of a few. Critics find utilitarianism unjust and disagree that the distribution of happiness is a fair measure of moral or ethical behavior.

Professional practice standards include the obligations and other standards of ethical conduct by professionals. The application of acceptable conduct is usually referred to as standards of "due care" or standard of care and is determined by the customary practices of professional community itself. Those without the requisite expertise are unqualified to determine what the standard should be, because they do not have the special knowledge that distinguishes the professional, as described in Chapter 2. Some of the difficulties with relying on professional standard of care as a guide to behavior is that the practice standard may not be suited for some circumstances where not enough information is known or the information relied on is inaccurate. There is a presumption in the standard of care model that professionals actually have sufficient knowledge and expertise to determine necessary precautions or standards of performance. It is equally likely that evaluations of due care are based not solely on objective professional judgments of facts but also on ethical ones. As knowledge moves forward on many fronts, how do the professions assure themselves that the standard of care reflects what is actually believed to be true in the natural world and in society? How might a change in the moral reasoning of society lead to changes in the standard of professional care? The standard of care would change to reflect this newer reasoning of course.

What Obligations Do We Have to Other Living Things?

The relationship between professional ethics and sustainability presents a number of challenges to the traditional approaches to moral reasoning. Among the more compelling issues are our obligations to the future and to other living things. Moral standards are not static in practice, and they change to reflect the changes in the culture.[6] Traditionally, moral standing or moral rights have been limited to people. Moral rights are essentially an entitlement to act or they have others act in a certain way. Although there are fundamental values common to all cultures, the particulars of moral behavior are a function of the values and circumstances of a given society at a given time and place. In this way, a belief in the sanctity of life, for example, might be expressed in different ways and at different times. Morally reasoned rights originated primarily in special relationships between people (e.g., craftsman–customer, teacher–student, parent–child, and neighbors), but a wider appreciation of the common interests of all people has led to wide recognition of human rights, which are not simply functions of relationships.

Human rights are universal (everyone, everywhere and at any time, has these rights), equal (no one person's rights are greater or lesser than another's), not transferable and cannot be relinquished, and natural rights (not derived from human institutions).[7] Negative human rights reflect the interests that human beings have from being free from outside interference. Positive human rights are the interests we have in receiving certain benefits (right to education, right to life, political participation, etc.). Any right a person has, he or she also has a correlative duty to act in a certain way. For example, you have a right to be fairly treated, but, in turn, you must treat others fairly. These rights can be expressed as negative rights (a person is enjoined from certain behaviors) and positive rights (a person has a duty to do something or behave in a certain way).

Until recently, little thought was given to extending moral standing to other living things or to the future. For the most part, such attention that was given viewed other living things primarily in terms of how they might best be used toward the interests of people. For example, in the First Treatise of

Government, Locke justifies the use and inferiority of plants and animals in the *Book of Genesis* and goes on for several pages listing the various rankings and uses.[8] In his lecture "Duties Toward Animals and Spirits," Kant said "Our duties toward animals are merely indirect duties towards humanity... If then any acts of animals are analogous to human acts and spring from the same principles, we have duties towards the animals because this we cultivate the corresponding duties towards human beings" and "Tender feelings towards dumb animals develop humane feelings towards mankind."[9]

In the seventeenth century, philosophers assessed the world in terms of primary and secondary qualities. Primary qualities consisted of the physical aspects of an object: its size, weight, material, how it moved, and similar characteristics that embodied the thing. Secondary qualities were those things that existed because of the interaction of the object and the observer. Characteristics such as color, taste, feel, and smell were functions of the interaction between observer and object; consequently, these secondary qualities were believed not to be entirely of the object itself. The role of science then was to describe the primary qualities of a thing. Secondary characteristics were not the purview of science.[10] In the intervening years, as science provided a clearer understanding, the lines between primary and secondary qualities blurred, so that, today, such distinctions are not routinely thought to exist.

The fundamental distinction between objectivity and subjectivity in Western philosophy is an important consideration in understanding the moral reasoning that supports an environmental ethic. Science studies the objective world, the real world, and is said to be "true," whereas the common interpretations of this world by individuals are subjective. The subjective world is one formed from perception and weighted by value and self-interest; subjective findings are not "true" in the sense of objective findings. Our subjective views, though, have been routinely challenged, as our understanding of the objective world improves. Moral standards are concerned with behavior that is of serious consequence to human welfare and accordingly that are significantly informed by our subjective views of the world. Ultimately, however, the soundness of a moral standard depends on the adequacy of the underlying reasons and hence is powerfully influenced by science.

Utilitarian Views of Nature

The view of Kant and other early philosophers that other living things are subject to the interests of humans continues even today. In his piece "People or Penguins," William F. Baxter presents a modern, utilitarian view of nature and the environment.[11] Baxter asserts that every environmental issue should be decided in favor of the greatest satisfaction of human interests based on the following four criteria[12]:

1. To discuss a problem, it is necessary to be able to describe the problem in objective terms.

2. There should be no waste; in his words, nothing should be "employed so as to yield less than [it] might yield in human satisfaction."
3. Every human should be regarded as an end, not as a mean, to the betterment of another person.
4. Every person should have an equal opportunity to improve his share of satisfactions.

The basis for this formulation is articulated in six points that are, in essence, a list of the difficulties in objectifying other living things or intrinsic environmental values and in ascribing human values to nonhuman objects.[13] The six reasons that Baxter gives are as follows:

1. People will always act in their own interests, because it is their nature, and will not act in another way.
2. Nature will be preserved because humans need it to be preserved. There is no "massive destruction of nonhuman flora and fauna" threatened.
3. What is good for humans is often good for "penguins and pine trees," so in this way, humans are "surrogates for plant and animal life."
4. There is uncertainty as to how any other system could be administered.
5. If nonhuman things are to be valued as ends rather than as means, it requires that someone determine how much each one counts and how they are to express their preferences.
6. These questions raise the question of what we ought to do, which in the end is uniquely human and meaningless to nonhumans.

Instead, Baxter argues that all of our actions should be assessed only in terms of his four criteria and the maximization of human satisfactions. His argument, however, contains several flaws. First, he requires that nonhuman subjects have an objectively assessed value as a precedent to moral consideration, but presumably, he could provide no such objective value for a human life. How would we value the human life—on the basis of the economic value of its chemical components or on the income earned or potentially earned in a life at work? What about the value of contributions to society of a great teacher, leader, or artist?

Second, he assumes that the environment and the economy are a zero-sum game and that a unit of work or resource expended there, by definition, requires an offsetting shortage somewhere else. He supposes that the costs associated with pollution control are necessarily expressed in "terms of other goods we will have to give up to do the job."[14] In fact, pollution controls create jobs and offset the externalities of no pollution control. Fairly robust economies have thrived without apparent offsetting negatives effects of pollution controls, as he suggested.

Third, Baxter assumes that there is some central authority that balances this real world. In fact, the world and our relationship to nature are far more complex than his argument recognizes. Science has found objective links between human interests and the environment that he does not account for. Fourth, he suggests that the only workable relationship with nature is one of subservience because of human nature. In fact, this is a problem of modern human society. Many other cultures do recognize our connectedness to nature and our duty to it. These cultures are successful and also human but use a different worldview from the one that Baxter assumes to be necessary. Last, Baxter's view of satisfaction does not account for human satisfaction in the future and assumes that the satisfaction of current generations is our only duty. In the end, the utilitarian view of our obligations to other living things, at least as argued by Baxter, is insufficient. We know, for example, that living and nonliving parts of our environment frequently have values that far outweigh economic or short-term human preferences. We understand that the environment is systemic and functions because of the interaction of many living and nonliving parts that Baxter's approach simply does not account for.

Speciesism

By the 1970s, our understanding of the interrelatedness of living things and the systemic fabric of life prompted some to begin to call for a greater moral inclusion and the recognition of the moral value of other living things, if not actual moral rights.[15] Singer's premise begins with a "fundamental presupposition" of the "basic moral principle" that all interests are entitled to equal consideration, that is, that any being with moral standing counts or matters. Singer writes, "the argument for extending the principle of equality beyond our own species is simple, so simple that it amounts to no more than a clear understanding of the nature of the principle of equal consideration of interests....our concern for others ought not to depend on what they are like, or what abilities they possess...[16]." In response, Des Jardins appropriately asks, "What characteristic qualifies a being for equal moral standing?"[17] Singer quotes Jeremy Bentham:

> "The day may come when the rest of the animal creation may acquire those rights which never could have been withholden from them but by the hand of tyranny. The French have already discovered that the blackness of the skin is no reason why a human being should be abandoned without redress to the caprice of a tormentor. It may one day come to be recognized that the number of legs, the villosity of the skin, or the termination of the *os sacrum*, are reasons equally insufficient for abandoning a sensitive being to the same fate. What else is it that should trace the insuperable line? Is it the faculty of reason, or perhaps the faculty of discourse? But a full grown horse or dog is beyond comparison

a more rational, as well as more conversable animal, than an infant of a day or a week, or even a month, old. But suppose they were otherwise, what would it avail? They question is not, Can they reason? Nor can they speak but, *Can they suffer?*"[18]

Singer then says, "if a being suffers, there can be no moral justification for refusing to take that suffering into consideration."[19] "If we make a distinction between animals and [severely intellectually disabled adults, infant orphans], how do we do it, other than on a basis of a morally indefensible preference for members of our own species?"[20] In the section "Speciesism in Practice," Singer goes on to discuss animals as food. He observes that people in environments in which animals must be killed for food are justified, but the rest of us who are not in such straits cannot morally justify taking animals for food. The cruelty of the treatment of animals in the conversion of calories and treatment of animals "like machines" are specifically enjoined in Singer's book. It is interesting, though, that he seems to be less concerned about free-range and meadow-raised livestock than about the factory farm. While I agree that the factory-farming practices are cruel and deserve our condemnation, I am not sure it is an argument for extending "rights" to livestock. Can Singer argue that a free-range chicken is being treated better than a wire-cage-raised chicken, as it is harvested? Were the evils of racism and sexism mitigated simply by eliminating slavery or chauvinism?

Speciesism is a term used by Singer as a corollary to sexism and racism.[21] He supposes that it is wrong to deny other species moral standing on the basis of species. Our anthropocentric bias according to Singer can be likened to race bias. Racial prejudice, however, is intraspecies prejudice rather than interspecies prejudice. The fact of the matter is that bias in the form of preference is morally permissible. It would be morally acceptable, even expected, that, all other things being equal, I would act to save my child, my sibling, my friend, my neighbor, my countryman, etc., in a sort of descending order of preference, in order of my bias. Clearly, all other things being equal, there should be an expected bias in favor of and in preference for our own species.

Ultimately, Singer must deal with what Baxter referred to as the administration of the system he proposes. How would such a world work? How would the interests of one species be balanced against the interests of another? Whose interests and rights would have greater weight, and what would the basis of those distinctions be? In the end, predator eats prey. Human beings are in effect managing the world, consciously or unconsciously. We have become so pervasive in our influence and effect that we must take our behavior into account. Morals are fundamentally based on human rather than individual interests. Morals are different than etiquette and law. An action may be legal but unethical/immoral, or an action may be illegal but morally right. The law can be seen as the floor of moral conduct; that is, the minimum acceptable level of morality and understanding of the origins of moral standards may be less important than whether they can be justified or not.

Most philosophers agree that waste is wrong. Baxter uses it as one of his four criteria. Singer's argument is at least partially rooted in the waste involved in our treatment of animals as food. Even Locke limited property to what could be held without spoilage and enjoins against waste: "*God has given us all things richly [sic]* is the Voice of Reason confirmed by Inspiration. But how far has he given it us? *To enjoy [sic].* As much as anyone can make use of to any advantage of life before it spoils; so much he may by his labour fix a Property in. Whatever is beyond this is more than his share and belongs to others. Nothing was made by God for man to spoil or destroy."[22]

We have begun to realize that we are indeed subject to our environment and the rules of nature, while at the same time, we are affecting the systems that compose nature. As moral animals, we must take responsibility for what we do. As intelligent animals, we must reflect on the objective that world science reveals to us and incorporate it into our subjective understanding. Further, perhaps the argument is not whether individual components of nature have rights but whether nature itself, as a system, does. If the absence of sentience is a concern, perhaps we can simply extend our moral umbrella to cover future generations of our species, who will require resources, and to environmental quality composed of diverse functioning ecosystems. Our moral duty is not to a single species but to the protection of nature itself. The value of a single species may be quite different when considered as part of an ecosystem as opposed to a unique and separate entity simply because we may not understand its relationships and roles within the system. The value to the system of nature may be incalculable, because we simply do not understand how this species interacts in its ecosystem in the long or short term.

What obligations do we have to other living things? In the end, moral consideration of other living things appears to be in order. Perhaps "the insuperable line" Bentham wonders about is the capacity for moral reasoning. Our capacity for reason requires more of us. We appoint advocates to protect the interests of those unable either by act (criminals) or by circumstance (children, insane, and intellectually disabled). Perhaps our knowledge requires a similar advocacy for nature and the future.

Since this was first written, two notable instances have occurred regarding animal rights. In the first, an animal rights organization filed suit on behalf of a macaque called "Naruto" to have the copyrights to a photograph claimed by David Slater, a British photographer. Slater had set up his camera in the area and provided for the macaques to possibly activate the camera while he remained at some distance. A resulting photograph was widely used in the media without attribution or payment to Slater. When he made attempts to claim compensation, the suit was filed on behalf of Naruto as the proper owner of the photograph.

The U.S. Copyright Office issued a rule that copyrights could be issued only to human beings; the resulting suit claimed that the original law authorizing copyrights did not mention species. A U.S. District Court Judge ruled in favor of Mr. Slater. The animal rights group appealed and eventually the plaintiffs and Mr. Slate settled out of court. Slater promised that a portion of the proceeds from the photograph would be donated to the protection of crested macaques. A joint statement declared that both parties "agree that this case raises important, cutting-edge issues about expanding legal rights for non-human animals, a goal that they both support, and they will continue their respective work to achieve this goal."[23]

In another instance, Germany amended or revised its constitution to declare that animals deserved the right to have their dignity protected. A quick Google search will result in numerous efforts and attempts and some successes to extend rights that have traditionally been limited to humans to other species. So, the status of interspecies rights is not as firm as it might have once been supposed.

Who Owns the Environment?

The challenge of sustainability lies at least in part in our ability to resolve some of the underlying principles of our society's worldview. John Locke's view of private property is fundamental to modern society and economics. In a culture where the individual is so highly valued and property has become a measure of distinction, our environment has been discounted and significantly compromised, at least in part, as a result. Capitalism is based on private ownership of the means and output of production. Private property is fundamental to the concept and cannot be separated from it. While this treatment is not inherently in disagreement with these principles, these are concepts that inform our worldview and are, for the most part, unquestioned but have significant implications for our relationship with nature. Among the challenges we face is finding a way to reconcile the culture of individuality and the paradigm of property with an emerging recognition of a need for sustainability and intergenerational equity.

The idea of private property extends from the conclusion that if things found in nature are to be of use to a person, there must be some means of appropriating them before they can be of use. No one would act to improve a thing without first being sure that their effort would serve them in some way or that they would have a right to the product of their effort. John Locke conceptualized the right of property as beginning with each person's ownership of himself, his own body. By extension, then, if a person should act on something found in nature, then by the 'labour' of his body and the 'work' of his hands the thing becomes his.[24] The investment of labor increases the value of a thing, and by virtue of this investment in its value, the thing becomes the property of the worker. Locke describes that a ripe apple on the branch of a tree (assuming this is a wild apple tree, held in common) is the property of no one and has no value while sitting there. If a person should come along and pick it, he has invested the work of picking it; it has additional value of having been picked and is ready to eat. This work has earned him the right of ownership and disposition, in this case to eat it or to give it to another to eat. Locke says, "The labour that was mine, removing them out of the common state they were in, hath fixed my property in them."[25] Having increased its value by making it available, it is his property, and he may decide its disposition.

Locke observes, however, that there must be limits to property. "[T]he same law of Nature that does by this means give us property, does also bound that property too...As much as any one can make use to any advantage of life before it spoils, so much he may by his labour fix a property in. Whatever is beyond this is more than his share, and belongs to others. Nothing was made by God for man to spoil or destroy."[26] Locke limits this conservation however to things that would spoil. With the invention of money there came a way to store more then what one could use in the form of durable value (gold, gems, currency). With money a person might "heap up as much of these durable things as he pleased..."[27]

With the capacity to store wealth or work in money, the acquisitive nature referenced by Adam Smith could be expressed. Locke observed that "...it is plain that the consent of men have agreed to a disproportionate and unequal possession of the earth-I mean out of the bounds of society and compact; for in governments the laws regulate it; they having, by consent, found out and agreed in a way how a man may, rightfully and without injury, possess more than he himself can make use of by receiving gold and silver, which may continue long in a man's possession without decaying for the overplus, and agreeing those metals should have a value."

Based at least in part on these ideas, economists refer to things found in nature as free goods; that is, they have no value in their natural state. It is only when the apple is picked and made available that it has value—likewise, for oil or lumber or anything in nature. The oil company pays for equipment and labor and for access over the land, but when it removes the oil from the ground, it pays nothing for it; it is free. It is the sum of the acts of finding, drilling, pumping, refining, and distributing the oil that is represented in its price on the marketplace, not the intrinsic value of oil. Free goods are not to be confused with the concept of public goods, which are goods or services whose benefits are not limited to one person; if one benefits, everyone benefits. As implied by the term, public goods cannot belong to one person. Examples might be air or national defense.

This devaluation of things in nature is compounded by the concept of discounting in modern finance. Discounting is simply the belief that future value is probably going to be worth less than the current value, so value should be taken as soon as possible for the best possible return. This is especially true when dealing with natural resources. For example, an acre of trees to be logged will continue to grow, adding wood over time, but rather than waiting, the trees are taken as soon as they have reached a size worth the effort of the harvest. To wait is to risk disease, fire, interference from other competitors, introduction of new materials or technologies that changes the value of the lumber, or any other possible problem. Harvest the trees as soon as possible, and risk is mediated, and the value can be invested in ways that grow faster than trees. Future earnings will also be paid in currency that is worth less than it is today. To offset the devaluation in currency value, the trees have to grow even faster. In this scenario, the trees as they stand have no value, and the trees in the future represent less than no value because of inflation.

The implications of these underlying concepts of property, free goods, and discounting are clear. In the purely economic view, there is no value in the environment. Value is described in terms of rights to use property or in terms of value added to things taken from nature. The economic system does not readily acknowledge the systemic value of the environment. For example, wetlands were seen as useless, so in the past, farmers and public agencies had programs to drain and fill them. These activities were undertaken to make the land "productive," that is, available for use as farmland or for development. Such wetland literally sold for next to nothing in places and was given away by the government in others.

We understand the systemic role of wetlands in the life cycle of living things and in the purification of water and air. As it turns out, wetlands are among the most productive natural landscapes. The approximate values of wetlands in the Chesapeake Bay watershed were discussed earlier. So, for every acre lost, the people of the region, in essence, choose to pay additional costs for clean water or accept a corresponding loss in environmental quality. The effects of these costs are real but often difficult to assess in terms an individual might appreciate, and so, the value is seen as intangible, of no use in the marketplace. Still, the costs mount, and the implications add up.

The difficulty we face is ultimately that environmental values are usually not expressed in economic terms. The two systems, economy and environment, are treated as if they exist apart from one another, distinct, unaffected by one another, but, of course, they are not. Sustainability in the end is a means of conducting environmentally sound economic activity for socially desirable outcomes. Too often, environment and economy are considered as diametrically opposed to one another, a question of clean environment or healthy economy. There is nothing inherent in either, however, that precludes or excludes the other. The challenge lies in our definitions of quality. Economists are concerned because it is our economic behavior that must accommodate the change. Sustainability requires a long view. If we are to become sustainable, the scope of our economic considerations must extend to include our impact on the environment and on the future. In short, we must begin to incorporate the externalities of our economic activity into the costs and undertake some consideration of intergenerational equity. Externalities are those costs or benefits of a good or service that are not included within its price. A negative externality might be the costs of health care or lower property values imposed on a community by a polluting factory. Clearly, concepts such as free goods and discounting must be evaluated with some thought to the future as well as the short-term environmental impact. Accounting for the externalities of our actions is among the most basic principles of our sustainable future.

Nature as Property

The syntax of Locke's use of the word "property" speaks to a larger meaning than the contemporary use of the word. At the time Locke wrote the Second Treatise, the word "property" was understood to refer to right in something, but the thing itself did not have to be material. "A man could have a property in a piece of land, but he could also have it in his life and in his liberties," Theodore Steinberg wrote.[28] In the past property, owners were not said to own the land but to hold the land, hence the term freeholders. Landowners were said to have certain rights to the property, but in many instances, others also had rights to the same land: for example, land held in common or held under a feudal system, where a freeholder's rights were subject to another. In contemporary use, the term has come to mean the actual object, a piece of land or a thing. The property is the physical object, not merely rights relative to it. In common modern parlance, when I purchase a piece of property, I act as if I am actually buying the land, taking ownership of the physical ground rather than acquiring rights to use the land. Property rights, however, refer to the rights and obligations associated to the land rather than the land itself. I purchase the right to use or dispose of the land, as opposed to actually owning the land itself.

The difference is an important one. My rights are restricted in various ways. I cannot negatively affect my neighbor's enjoyment of his adjacent property, for example. Or my plans for the parcel may require agreement and approvals from local authorities. So, the privileges or rights of ownership are not absolute. My property rights are limited by the need to protect the health and welfare of the public and my neighbor's rights to enjoy his property. There are numerous other ways in which ownership is limited by the interest of the public and individuals. It is said in real estate that there are three conditions of value: location, location, and location. On a deeper level, though, it is the rights associated with the location that have the value, not the land itself. A piece of property that is sold with conditions that preclude the buyer from building, restricting access, or harvesting the land has little value, regardless of its location.

As a result, it is the prevailing view in law and public policy that no individual can claim to have been "harmed" by a loss of general environmental quality of a resource held in common, since no one owns it, no specific harm is experienced, and everyone is equally affected by the loss.[29] This view seems to stand logic on its head; however, why should anyone be able to appropriate an element of quality from a resource held in common and presumably deny its use of everyone else? Nature and the products and benefits of nature must be a public good; that is, no one person may own or exclude another from it. "As a good, environmental quality exhibits joint nonvoluntary transfer and nonexcludable rights—traditionally described as the two characteristics of a public good."[30] Under such a vision, it could be argued that even a private property owner would not have acquired the right to diminish environmental quality, a public good, as part of the title of a parcel of land. However, little can be done with the rights to property, without at least some nominal environmental effect. Without some privilege, the rights of private property would mean nothing. The difficulty is in knowing the value of the impact and the cost of not acting to protect nature.

The Value of Property and Nature

To consider the environmental impacts of the use and development land without consideration of the economic implications is a pointless effort. In the end, sustainability will be expressed as a balance of values. To date, only the economic values have been considered. As we go forward, it is necessary to capture the economic and environmental externalities, so that this balance can be achieved. It is critical to understand both sides of the sustainability ledger if our efforts are to succeed.

Land development in the United States represents considerable economic value. Housing real estate value less mortgage expense was estimated to be about $18 trillion in 2006, and while there have been losses estimated to be around 20% in the years that followed, this still represents the largest pool of personal wealth in the United States.[31] Commercial real estate holdings are valued at about $5.3 trillion.[32] Public policy is predictably geared to protect and encourage this accumulation of wealth. Associated with the value of real estate is the role that development plays in the gross domestic product (GDP). Construction alone contributes between 4% and 5% of the annual GDP of the United States. There is considerable economic investment and equity in real estate and development. Real estate taxes, transfer and development taxes, are major sources of local and state revenues. Our understanding of property valuation within a marketplace is fairly sophisticated, and we often hear issues and proposals equated to us in terms of negative or positive impacts on property values.

In light of this, how are we to find comparable value in nature? In 1973, economist Colin Clark demonstrated that by using standard economic models, the best economic choice for whalers was to hunt the blue whale to extinction as quickly as possible. This choice would provide the whalers, and by extension, the human race, the greatest possible return. The cost of

Is There an Ethical Obligation to Act Sustainably?

limiting harvests until blue whale populations recovered and then hunting only at a sustainable level would yield less profit than killing them off and investing the profit. Clark thought that there was an unmistakable flaw in the model. E.O. Wilson recounts Clark's argument: "The dollars-and-cents value of the blue whale was based only on the measures relevant to the existing market—that is, on the going price per unit weight of whale oil and meat. There are many other values, destined to grow along without knowledge... as science, medicine and aesthetics grow and strengthen, in dimensions and magnitudes still unforeseen."[33] Wilson and others raise the question as to whether contemporary economic models are able to evaluate or express the value of nature, since they do not account for future value or current value beyond the utility of the products from a dead blue whale. The models all discount future value, as described earlier.

It is not that economists cannot put a value on natural resources. There are many models for the valuation of iron ore not yet extracted or oil yet to be pumped. The reason for the inability of economic models to consider or account for the value of other resources is precisely that those resources are not traded. It is the marketplace, the willingness to pay, that establishes value. The concept of willingness to pay is a measure of how desirable a thing is to possess, to own. What was the value of a blue whale in 1970 in the ocean? Zero. Nothing. What is not possessed has no value. What might the value of the same whale be in 100 years? 300 years? Difficult to say.

To identify the value of a thing in nature, unpossessed and unowned, other means of assessing value are necessary. Willingness to pay is a fundamental economic concept that is essentially a measure of the perceived utility of good or service, in other words, one way to impute the value of an unpriced good or service is to determine how much someone is willing to pay for it.[34] If it was as simple as this, however, the model of discounting future value could not be refuted, since only market utility is measured. Sagoff suggests that the model needs to be reassessed: willingness to pay "is not a value or a definition of value or a reason to value anything...To find out what we are willing to pay for, we have to determine what we value, not the other way around."[35]

We do find intrinsic value in nature that does not translate simply to market value. Value as a resource yet to be considered, such as systemic values, beauty, and solace, comes to mind. If these things are of value to us, willingness to pay is simply an inadequate measure and incapable of expressing value.[36] We understand nature's value to transcend simple economic expression, but we are frequently faced with the proposition of tangible economic impacts weighed against transcendent values of nature. The economic values of nature are real but difficult to assess and to make real in the same way as real estate values.

Reconciling Private Property and Sustainability

Capitalism is based on private property because private property is valued in the marketplace. Our current understanding is that the right to property ownership is a fundamental extension of our claim to individuality.[37] All things within this system, however, are valued according to their utility or the willingness of a person to pay, but we recognize that all things valued cannot necessarily be expressed in those terms. To account for nature in a sustainable transaction will require different models and new considerations. The capacity to develop those models, however, will ultimately be a function of which theory of justice is chosen: utilitarian, libertarian, or Rawlsian.

Utilitarians will have to consider the greatest utility in terms of the environmental consequences of behavior and decisions and whether such long-term considerations can outweigh more short-term satisfaction. Can the satisfaction of the unborn be balanced against the living? Libertarians must consider whether the rights of future generations represent an encroachment on the rights of the current generation. Rawlsian justice must fold the interests of future into those considerations within the veil of ignorance. If the interests of the future and the intrinsic interests of the earth are considered, how will they be accommodated within these schools of thought?

There have been a number of efforts to describe nonmarket environmental values.[38] These methods range from calculating the systemic value if specific environmental functions (e.g., wetland capacity to clean water) to future resource value (e.g., biotechnology value of unidentified species) and indirect value of natural resources (e.g., tourism value of forests or rivers). Interest in these proposals has been largely academic, because there is insufficient public interest or awareness to compel the change.

Edward O. Wilson has written an important book that calls for consilience among the various professions and disciplines. In Wilson's book, consilience is "units and processes of a discipline that conform with solidly verified knowledge in other disciplines have proven consistently superior in theory and practice to units and processes that do not conform."[39] In other words, one discipline cannot ignore what has been proven to be true and to work in another. We must act on what we know to be true. If we have a moral duty to assist and prevent harm, then among professionals, consilience may be obligatory. Design is by its very nature an interdisciplinary process. Designers must consider many aspects of a project in the design synthesis; it is a consilient process.

Weak sustainability allows impact to be exchanged for equity, but that implies value that might not exist in the future. Strictly speaking, weak sustainability may be difficult to support.[40] Though it appeals to economists, it does not pass the tests of good science and is inconsistent with what we know to be true about the natural world and the long-term interests of humanity. Economics as it is currently formulated is based on activities conducted in isolation from their impacts and extended costs.

Strong sustainability, however, limits our ability to act at all. Since we cannot know the future, we cannot risk any action that would have an impact. In this case, the present is always a hostage of a future that doesn't even exist.

Clearly, neither argument is entirely persuasive. Actions must be made after balanced considerations. The difficulty does not lie, however, in reconciling science and economics. The difficulty lies in understanding our values and putting them into practice. Design professionals have important roles to play in the process. Few are better prepared to understand the implications of their acts and to formulate effective alternatives. Fewer still have greater direct influence through their work on the quality of the environment and the sustainability of our society.

Do No Harm

It has been suggested that because of the impact designers have on our quality of life, the environment, and the future, they should be subject to the same kind of professional oath as physicians. A number of forms for such an oath for design professionals have been proposed.[41,42,43] At the core of each proposal are the well-known do-no-harm words from Hippocrates "…I will use treatment to help the sick according to my ability and judgment, but never with a view to injury and wrongdoing."[44] Similarly, many would agree that the words of Aldo Leopold are apropos for designers of all sorts, "A thing is right when it tends to preserve the integrity, stability, and beauty of the biotic community. It is wrong when it tends otherwise."[45] Calls for an oath to "do no harm" have received little meaningful attention from the professions in general, and it would seem that such a step would be difficult in practice or for the professions to enforce and so would do little good.

Designers should be held to a higher standard of care than ordinary citizens, but the nature of the design professions is such that they differ in important ways from the classic ideal of a physician acting as an autonomous practitioner making a version of the Hippocratic Oath problematic. Engineers provide services directed toward things such as buildings, products, and infrastructure, whereas lawyers and doctors direct their attention toward individuals, usually clients. Designers' relationships to clients are indirect in that the services are directed elsewhere.[46] They must balance the interests of clients, community, economics, the environment, end users, and regulators to synthesize a design. Requiring the designer to do no harm would raise the question as to who defines harm?

All parties impacted by a project have different concerns and would weigh costs and benefits accordingly. Is the designer to act as sole arbiter in this process? Ultimately, the very idea of autonomy is called into question. If the engineer does not act with individual autonomy, can there be overarching professional autonomy? The Oath of a person without the autonomy to act is meaningless. In such a case, is the public's confidence and sanction warranted? It is unlikely that an oath to do no harm would have significant weight or influence under these conditions.

Role of Professional Standards

Professional practice standards are determined by customary practices of a professional community. The standard of care or due care is determined as what a reasonable qualified professional would do under the same circumstances and with the same information. Those without the required expertise are unable and unqualified to determine or make judgments as to what the standard should be. This is why doctors, engineers, architects, and the like are asked to testify when a colleague is accused of negligence or a lapse in the standard of care. Standards of care are not objective or definitive in the sense that they can be written down or codified beyond the codes of ethics or professional responsibility articulated by the various professional organizations.

The standard of care is always changing. New information on materials, for example, will lead to a change in what a designer specifies and how it might be used. The professional has a duty to maintain a working knowledge of those areas in which he practices, and clients and the public at large have a right to rely on this obligation. As the information available to the designer changes, it is expected that practices and the standard of care will change accordingly.

The standard of care for professional practice is fluid. For example, the standard may be too great for some circumstances where not enough information is known. When using new techniques or materials, how can professionals be certain that they actually have sufficient knowledge and expertise to determine necessary precautions or standards of performance? The professional relies on the duty to preserve health and safety in determining the appropriate care in such circumstances. In the end, decisions about due care are not merely professional technical judgments but ethical ones. Again, the professional is guided by values, in this case the values outlined by the professional community and the representative professional association. As the designer undertakes a design that is intended to meet some degree of sustainability, but that has no objective standard on which to rely, what will guide the process? While there is objective knowledge of human impacts on the environment and of examples of design that has minimized such impacts, there is no objective "toolbox" on which the designer may rely. The designer will seek to operationalize the underlying values of sustainability, as she understands them, and these are primarily value judgments, that is, ethical judgments. In the end, there is an ethical obligation for designers to take issues of sustainability into consideration in their work.

Endnotes

[1] Berry, Thomas, *The Dream of the Earth*, Sierra Club Books, San Francisco, CA, 1988, p. 52.

[2] Kempton ibid, p. 74.

[3] Beaucamp, Tom L., Norman Bowie, *Ethical Theory and Business*, Prentice-Hall, Upper Saddle, NJ, 1997, p. 34.

[4] McMahan, Jeff, "Moral intuition," *The Blackwell Guide to Ethical Theory*, Hugh LaFollette, editors, Blackwell Publishers, Malden, MA, 2000, p. 93.

[5] For the sake of time and space, I am going to limit my comments to the prevailing culture derived from Western Europe and the western Judeo-Christian experience. It should be noted however that many, in fact most, other cultures, existing and past, did incorporate nature. Nature in these cultures was inseparable from the culture, the day to day life, the very idea of being a person. This short paper will limit itself to the dominant world culture and its recent awareness of the need to accommodate the laws of nature. Also I will not provide the laundry list of environmental problems and challenges that serve to increase our awareness, though the list is compelling.

[6] Moral rules are not static. Our judgment reflects our experience and our interests. As conditions change and as we learn as individuals and then as cultures, our moral standards change to reflect our new circumstance. In the past slavery, institutionalized class systems and cruelty to animals were accepted as moral, but over time the moral barometer changed as the interests of the society changed. This expanding moral umbrella is a function of the security and stability of the human species and cultures. The scope of inclusion expands to cover more people and living things as our fates intermingle. Our moral vision includes others, at least in part, as a function of our ability to meet our own needs first (Maslow's Hierarchy of Needs). In the absence of those needs our social scope of moral inclusivity shrinks. Likewise, as our needs are met and secured, the aperture of moral standards opens.

[7] Barcalow, Emmett, *Moral Philosophy: Theories & Issues*, Wadsworth Publishing Co., New York, 1998, 2nd edition, pp. 202–209.

[8] Locke, John, *Two Treatises of Government*, Cambridge University Press, Cambridge, UK, 2000, pp. Chp IV pp. 156–177.

[9] Kant, Immanuel, *Lectures on Ethics*, lectures given between 1775 and 1780 transcribed by three students and edited by Paul Menzer in 1924. This edition was translated by Louis Infield, Hackett publishing Co., Chicago, 1980, pp. 239–240.

[10] Des Jardins, Joseph, *Environmental Ethics, An Introduction to Environmental Philosophy*, Wadsworth Publishing Co., New York, 3rd edition, 2001, p. 221. Tertiary qualities such as beauty, inspiration etc. were also discussed by early philosophers.

[11] Baxter, William F., "People or penguins," published in *Moral Issues in Business*, 8th edition, William H. Shaw and Vincent Barry, editors, Wadsworth Publishing, Belmont, CA, 2001, pp. 580–584.
[12] ibid.
[13] ibid.
[14] ibid, p. 583.
[15] Stone, Christopher D., *Earth and Other Ethics the Case for Moral Pluralism*, Harper & Row Publishers, New York, 1988, pp. 73–83.
[16] Singer, ibid, p. 52.
[17] Des Jardins, ibid, p. 114.
[18] Bentham, Jeremy, *The Principles of Morals and Legislation*, Prometheus Books, Amherst, NY, 1988, originally published in 1781, footnote to paragraph 4, Section I of Chapter XVII), singer quotes Bentham on Page 56 of his book.
[19] Singer, ibid, p. 59.
[20] Singer, ibid, p. 60.
[21] Singer, Peter, *Practical Ethics*, 2nd edition, Cambridge University Press, Cambridge, UK, 1993, p. 55.
[22] Locke, ibid. Chp V, 5–10, p. 290.
[23] The Guardian, https://www.independent.co.uk/news/world/americas/monkey-selfie-david-slater-photographer-peta-copyright-image-camera-wildlife-personalities-macaques-a7941806.html Accessed June 24, 2018.
[24] An Essay concerning the true original, extent and end of Civil Government (1690) from http://odur.let.rug.nk/~usa/D/1651-1700/locke/ECCG/govern02.htm, p. 26.
[25] ibid, p. 27.
[26] ibid, p. 31.
[27] ibid, p. 46.
[28] Steinberg, Theodore, *Slide Mountain or the Folly of Owning Nature*, University of California Press, 1995, p. 11.
[29] Brubaker, Elizabeth, "The ecological implications of establishing property rights in Atlantic fisheries," *Environment Probe*, April 1996.
[30] Gillroy, John Martin, *Justice and Nature, Kantian Philosophy, Environment Policy & The Law*, Georgetown University Press, Washington, DC, 2000, p. 40.
[31] Poole. William, "Real estate in the United States" remarks made to the industrial asset management council convention, 1/9/2007, http://www.stlouisfed.org/news/speeches/2007/10_09_07.html
[32] Standard and Poores Commercial Real Estate Indices, 12/31/2007, http://www2.standardandpoors.com/spf/pdf/index/SP_GRA_Commercial_Real_Estate_Indices_Factsheet.pdf
[33] Wilson, Edward Osborne, "What is nature worth?" *Wilson Quarterly*, 26(1), 280, 2002.
[34] Turner, R. Kerry, David Pearce, Ian Bateman, *Environmental Economics, an Elementary Introduction*, Johns Hopkins Press, Baltimore, MD, 1993, p. 108.
[35] Sagoff, Mark quoted in Campos, Daniel G., "Assessing the value of nature: A transactional approach," *Environmental Ethics*, 24(1), 71, 2002.
[36] Campos, Daniel G., "Assessing the value of nature: A transactional approach," *Environmental Ethics*, 24(1), 71, 2002.

[37] The moral justification of capitalism is interesting, particularly with regard to our assumption that we assume it is moral simply because it is so familiar to us. This is an interesting element but beyond the scope of this inquiry.
[38] Turner et al., ibid, p. 111.
[39] Wilson, Edward O., *Consilience*, Alfred Knopf, New York, 1998, p. 198.
[40] Ayres, Robert U., Jeroen C. J. M. van den Bergh, and John Gowdy, "Strong versus weak sustainability: Economics, natural sciences and "consilience," *Environmental Ethics*, 23(2), 166, 2001.
[41] An Engineer's Hippocratic Oath, Ch. Susskind, *Understanding Technology*, The Johns Hopkins University Press, Baltimore, MD, 1973, p. 118.
[42] Obligation of an Engineer, University of Minnesota Duluth, College of Science and Engineering, http://www.d.umn.edu/cse/ooerc/oath.html
[43] Hippocratic Oath for Scientists, by Nicholas Albery, http://www.globaliseasbank.org/BOV/BV-381.HTML
[44] Hippocrates, as translated by W.H. Jones, Harvard University Press, 1923 as reprinted in *Ethics, Tools and the Engineer* by Raymond Spier, CRC Press, New York, 2001.
[45] Leopold, Aldo, " The land ethic," as quoted in The philosophical foundations of Aldo Leopold's land ethic by Ernest Partridge. http://www.igc.org/gadfly/papers/;eopold.html
[46] Ladd, John, "Collective and individual moral responsibility in engineering: Some questions," in *Ethical Issues in Engineering*, Johnson, Deborah, editor, Prentice-Hall, Englewood Cliffs, NJ, 1991, pp. 26–39.

4
The Design Professional and Organizations

Education is a progressive discovery of our own ignorance.

Will Durant

Balancing Obligation and Opportunity

Most design professionals tend to work within an organization and are subject to a bureaucracy of some form. For the most part, they do not practice as a solo professional, which is the traditional model of other professions such as medicine and law. The core principles of professionalism are challenged in the real world by virtue of the designer's place within organizations. Can the designer act as an autonomous professional from within the ranks of a bureaucracy? Where does social responsibility fall in relation to the goals and objectives of the organization? Moral responsibility, like design, is forward looking. How do organizations and individuals balance their duties to employees and employers with ethical obligations to clients and public welfare? It is about what we ought to do to prevent harm and to promote good.

McFarland applies the Kew Gardens argument for professional responsibility.[1] Using the Kew Gardens principle[2] described by Simon et al., the moral minimum standard of behavior would include an obligation to respond to a person needing help under the following conditions: need, proximity, capability, and last resort. The Kew Gardens argument recalls the incident in which a woman named Kitty Genovese was murdered, while her neighbors ignored her pleas for help. The argument is made in four steps:

1. A critical need (a right or some good is threatened or at risk).
2. Proximity: in the area or in the case of engineering responsibility a function of notice or awareness.
3. Ability to help, constrained only but not risking damage to oneself or duties owed to others.
4. Absence of other source of help.

73

Need is necessarily qualitative and difficult to describe in advance, but as need increases relative to the other conditions, the affirmative duty to respond increases as well. To be so obliged, one must be in the proximity of the event or circumstance and be in a position to do something. Proximity and capability do not require unreasonable acts; however, as need increases and if our intervention is indeed the last resort, our obligation to respond in some way increases. The Kew Gardens principle is easily extrapolated to design professions. By virtue of special training and knowledge, such professionals may be among the few able to identify the character and degree of risk necessary to assess the need and, having made such an assessment, an affirmative duty to act on their findings to prevent harming others. Indeed, special knowledge may enable the professional to anticipate need in the design phase before it actually exists as a condition in the real world.

The obligation to protect public safety is an affirmative duty of the design professional. In addition to the affirmative duty to prevent harm or assist a person in need, however, there is an injunction against causing harm. Each person has a fundamental right not to be harmed by the acts of others. With every right comes a duty for others to not violate that right. These correlative duties for any right are (1) the duty to avoid depriving the person of their right, (2) to help protect the person from such deprivation, and (3) to aid the person who has been deprived.[3] Donaldson observes that the last may or may not be a duty as a function of circumstance. It follows then that the design professionals have an affirmative duty to prevent harming others and, by virtue of their training and expertise, a duty to make a reasonable effort to anticipate what harm might come from their work. The expectation that the designers must anticipate the consequences of their decisions leads to ethical duties to protect the environment and the interests of the future to the extent these can be foreseen.

Designers are the technical experts, and the public/concerned parties trust them to do the right thing. McFarland observes, however, that history has repeatedly demonstrated that science and technological systems are never fully characterized and understood. There is a degree of complexity in systems that escapes even the most rigorous characterization and understanding, and the degree of complexity associated with environmental impacts is significant. It is this sort of uncertainty that inspires the caution of Spier and Florman. How can the public be assured that its interests are considered and that the expected standard of care is provided by professionals operating from within an organization?

Design professionals working in the private sector have to be concerned with the interests of clients. As employees, they have obligations to the organization that employs them, and as professionals, they have obligations to the client and to the general interest of the public. Individuals routinely find ways to balance these competing obligations with their professional responsibilities. In many ways, the designer in this position is acting as a manager more so than a designer. Day-to-day practicality would encourage, within the limits of a professional standard of care, design professionals to tend to give the interests of their client or employer the greatest weight in their consideration, just as managers tend to give the greatest weight to the interests of stockholders. The managers, however, have a scope of concern and care limited to the interests of stockholders, and future considerations are most commonly held to the test of stockholders' interest. Concerns beyond the organization are limited. This narrower scope of obligations is quite different than the standard of care expected from a professional who has a duty to others beyond the client and employer.

In recent years, the interests of stakeholders have become more important in public and private organizations. For our purposes, "A stakeholder in an organization is any group or individual who can affect or is affected by the achievement of the organization's objectives."[4] Freeman constructs a Rawlsian argument for extending the traditional fiduciary relationship between managers and stockholders to include stakeholders.[5] John Rawls's theory of justice employs what he calls a veil of ignorance, in which participants in making decisions do so without an awareness of their own roles in the circumstances or their decisions. In other words, justice would be achieved if we were to craft rules that we would find just regardless of our circumstances in the application of those rules. In Freeman's view, if stockholders, managers, and society were to construct the rules of business from within a Rawlsian veil of ignorance, business would be conducted according to what he calls the Doctrine of Fair Contracts (Table 4.1). It would seem that if the design professions are to be held to a standard of care that includes sustainable thinking, it is plausible that a similar approach will be taken with

TABLE 4.1

Freeman's Doctrine of Fair Contracts

1. Principle of entry and exit: contracts must have clearly defined parameters of entry, exit, and renegotiation.
2. Principle of governance: rules for changing the rules.
3. Principle of externalities: any party to which a cost is imposed by a contract is, by definition, a party to the contract.
4. Principle of contracting costs: all parties to a contract must share in the cost of contracting.
5. Agency principle: any agent must serve the interests of all stakeholders.
6. Principle of limited immortality: management shall proceed as if it will continue to serve all stakeholders throughout time.
Recognizing that much in the way of enabling legislation would be required to effect these changes, Freeman proposes three principles to guide the reformation process:
1. Stakeholder enabling principle: corporations to be managed in the interest of all stakeholders.
2. Principle of director responsibility: standard of care for directors of corporations to observe the stakeholder enabling principle.
3. Principle of stakeholder recourse: directors subject to action by stakeholders if they fail in their responsibility or standard of care.

general sorts of design decisions and will guide the interactions that occur between stakeholders in a sustainable design process.

Would such a circumstance work against the professionals' autonomy, resulting in design by committee? The professions have historically been closely tied to the interests of clients and employers, but one could argue no more so than the manager of a corporation to the interests of stockholders. Goodpaster's view of the stakeholder relationship to management and the corporation is expressed best as a new synthesis of relationships. Stockholders cannot expect managers to act in a fashion that is contrary to what they could reasonably expect of any person in the community. He calls this the Nemo Dat Principles (NDP) after a Latin proverb *nemo dat quod non habet* or "nobody gives what he doesn't have," or as Goodpaster restates it, "No one can expect of an agent behavior that is ethically less responsible than he would expect of himself. I cannot (ethically) hire to have done on my behalf something that I would not (ethically) do myself."[6] The argument for management's consideration of stakeholder's interests is at least as valid when applied to the design professional whose work is likely to reverberate at least as much as a given manager. The design professional, however, must be expected to act as the gatekeeper, as dictated by his professional knowledge and ethical obligations.

There is a great need for the design professions to participate in the emerging practices of sustainability, and it is anticipated that they will adapt to the changes that society expects of them. The question for the professions is whether they will lead the process of change or be directed to do so. As discussed in Chapter 2, several professional groups have adopted the principles of sustainability into their codes of ethics, and changes in the built environment can be seen to a limited extent. If the design professions are to provide leadership and adapt to the demands of sustainability, the standard of care must embrace a stakeholder-based approach to all design and the interests of the future must be considered in that process. Expanding the professions' standard of care to include the interests of stakeholders beyond the professional's employer and the client and to include the concerns of the future generations would be consistent with the spirit of the traditional canons of the design professions and would serve the interests of sustainability. Ultimately, it may be the only way to fairly balance the design professional's responsibilities and obligations and the diverse interests of stakeholders in an emerging culture of sustainability.

Can There Be Deeds Without Doers?

The fact remains that the built environment and products of manufacture need to reflect concern with environmental quality, and design must be a clear voice of leadership and change. Actions that would speak louder than the statements in various codes of ethics pale by comparison. In the contemporary marketplace, the typical design professional, however, is an employee, not a free agent operating in isolation. If we are all acting as ethical autonomous professionals, how do we explain the environment we have designed and overseen the construction of? Of course, we cannot lay the problem at the feet of any individual person or designer. It is always the same problem described by Garret Harding in the "Tragedy of the Commons." The environment is our commons, and no one is to blame for its problems. We are all responsible. Design professionals may bear a greater responsibility born of special knowledge and a dynamic standard of care. Most design professionals, however, do not act as individual autonomous agents. Is a person operating within an organization able to make an ethical choice within his or her organizational role? And if so, is he responsible for doing so? For purposes of this discussion, ethical obligation and legal liability must be distinguished from one another. We must presume for purposes of discussion that the actions of individuals and organizations are legal.

Organizations develop and rely on an internal bureaucracy. They function to some degree on the predictability and limited scope of internal decisions made by the bureaucracy.[7] The nature of bureaucracy is to limit the authority and discretion of any one person within the organization. It is the systemic limits of this bureaucracy that may inhibit ethical decision making by constraining or limiting at least one of the four knowledge areas said to be necessary for moral choice.[8] This chapter explores the assertion that organizations, by design of bureaucracy or expert systems, can create outcomes for which no one is responsible. It is necessary to distinguish between the person as an ethical agent acting alone and as an ethical agent able to act within an organization. Based on that distinction, it might be possible to see if claims of ethical innocence from within an organization can be justified and if, indeed, there can be deeds without doers.

Ethical Agency

People working in modern organizations act within the parameters of the organization; that is, they act as part of the system,[9] that is, the organization. Any given person within an organization may be defined by their scope of authority and responsibility. Typically, authority is closely held and tends to be concentrated hierarchically in the formal leadership of the organization, whereas responsibility is disseminated by the organization and pushed down into the lower echelons to the degree that it is possible and desirable to do so.[10] Organizations operate on the expectation that each person, and by extension each group of people, will complete their duties and responsibilities in a manner that supports the objectives of the organization. Performance is evaluated based on the expectations created by the organization's objectives and the degree or effectiveness with which an individual or a group makes specific contributions toward those expectations. For example, creativity is generally included in most performance appraisal systems and is considered valuable, but creativity is generally defined only in the terms of the organization's needs and goals. Creative contributions are those that increase productivity, provide valuable distinction to the firm, or contribute to the bottom line, either directly or indirectly. Where someone working in research and design may be *required* to be creative, another person working in the same organization, say in document production or drafting, may be *required* to never deviate from a specification. Creativity in the latter role is a narrowly defined characteristic.

Like creativity, responsibility is apportioned differently throughout an organization. A person in the lower echelons of an organization may not see how their actions fit into the larger scheme of corporate objectives or outcomes. Even if they were to have knowledge, they have neither the authority nor capability to influence the outcome. For the design professional to act ethically, he must be in a position that allows him the ability to recognize and understand the circumstances of choice and the capability to respond to the ethical requirements of the situation.[11] A person acting in a line function may or may not have the knowledge but certainly will not have the capability to alter the behavior of an organization. Clearly, that person cannot act as a moral or ethical agent in his or her role as an employee. The scope of authority expands and the understanding of outcomes becomes clearer as a person moves to positions of greater authority and responsibility.

Individuals do not, in actuality, work in isolation from the facts of their work. If a line person does have knowledge that his actions lead to an outcome believed to be unethical or will cause harm, his choice is clear, to continue or not to continue. He may decide that the outcome of his actions, though indirect, is so onerous that he cannot continue. Alternately, if the job provides the income to his family and there are no other options for employment or income, his first duty may be to continue to work. He may choose to work for change within the organization or continue to look for alternatives to employment or merely accept his plight, but his actions are limited to the scope of his knowledge and capability. This choice, however, is a personal choice, with little direct influence on the organization itself or on the eventual outcomes resulting from his work.

The moral character of an organization is established by the leaders.[12] The manager, while still a functionary of the organization system, has a broader range of knowledge and responsibility. The manager has more capability to influence the behavior of the organization. He is also more likely to be able to meet the four knowledge conditions necessary to make an informed moral choice. The manager who chooses to accept organizational inertia, that is, to "go-along-to get-along" rather than challenge, is making a choice, albeit, not necessarily a moral one. Robert Jackall offers his observations of behavior in organizations as an illustration of a bureaucratic culture of uncertainty and fealty. In this culture, success and failure in the organization are the result of random, often discontinuous, events over which the individual has little influence and no control.[13] We would have to believe that success in this environment is not earned; rather, it is provided for those of us who are most agreeable (or least disagreeable), have a good sense of corporate fashion, are deft at avoiding decisions, and work for people who, like us, are going to be provided success by their betters, regardless of merit. While it would be naïve to suggest that these things are never true, it is also a gross simplification to suggest that they are always true. In Jackall's organization, change could never occur because there could be no reward great enough to offset the risk of failure, but business organizations must change constantly. That they do change, and that some organizations are often opportunistic and reactive, stands in contrast to the bureaucratic culture observed by Jackall.

This capability to change also speaks to some degree of the organization's capacity to act as a moral agent. Within the organization are managers or others who have the responsibility and opportunity to chart the course of the organization and to determine the vision and strategies of the organization. It is these individuals who establish the goal and objectives and the policies and, ultimately, the behavior of the organization. Power is concentrated in the hands of these people. Though responsibility is pushed down into the organization, power cannot be entirely divorced from responsibility. Within the organization are individuals who are given the authority and the responsibility for the outcomes and methods of the organization; these people have the capability to change the organization and meet all four of the knowledge conditions. These are the doers.

The capability to act as a moral agent on behalf of an organization is not fixed at one level in the organization. Supervisors and employees are able to make ethical choices every day but usually only within the scope of their assignments. Some people within the bureaucracy may act as moral agents for the organization within the context of their authority. At such times, it is not surprising that individuals might retreat into the bureaucracy for protection.

To act as a moral agent, four knowledge conditions must be met and the person must be free to act or decide. In organizations, the knowledge necessary to understand outcomes and actions and the power to change the behavior of the organization may be concentrated in the hands of a few. Such leaders establish the flavor of the organization's behavior. Moral choices are made by organizations at various levels and by individuals within the scope of their authority, but these are usually within the operating discretion of the individual. Moral responsibility for the organization must lie first with those who have discretion and knowledge to act but also with those who determine how the individual will be evaluated within the organization.

The Descent of Corporate Obligation

Professional firms tend to be organized as partnerships or some type of corporation, but for purposes of our discussion, the form an organization takes is somewhat less important than how it behaves. Corporations[14] were first formed as municipal organizations, religious orders, or universities, and in this early form, they were distinguished primarily in the distinction of the organization from its members. The corporate entity allowed the individual or group of individuals a limit to their personal liability and so provided some protection to persons serving in local government or institutions. It was not until the late sixteenth and early seventeenth centuries that business entities were formed as corporations under English law.[15] English law recognizes two main corporate forms: corporations aggregate and corporations sole. The former is composed of groups of "natural persons," while the latter includes "bodies politic constituted in a single person." The corporate sole usually consists of a person such as a bishop or a dean who has power by virtue of an office to hold property. Example might be the bishop who controls property and authority by virtue of his office; when he should pass on, the power and property are retained by the office and his successor. The corporate sole was the earliest form of corporation.

Eventually, the corporate aggregate became more common, as individuals used capital for investments but wished to limit liability and to pool resources. The corporation aggregate has rights and duties as if it were a natural person separate and distinct from its individual members. As such, it is described as a "juristic person"[16] who is endowed with legal personality, with rights to property and the like. The corporation, however, has an existence beyond the life of any individual or group of individual members. The corporation allows members or stockholders to invest capital in a venture and limit their liability to only their total of their investment. According to Shaw and Barry, "Limited liability is a key feature of the modern corporation. It means that the members of the corporation are financially liable for corporate debts only up to the extent of their investments."[17]

Early business corporations were commonly formed around a grant or trading monopoly that attracted private capital to the development and exploitation of the New World. While members' liability may have been limited in the early corporations, it was not necessarily combined or pooled. Members of the corporation often financed individual projects under the corporate umbrella, as it were. Under such an arrangement, if the venture failed, only that member's capital was at risk. As the scale of ventures increased, however, members pooled capital and shared risk and liability.[18]

In 1766, Adam Smith completed *An Inquiry into the Nature and Causes of the Wealth of Nations*,[19] in which he observed that open and free competition serves and regulates a community by relying on the self-interest of its members. When there is equal access (laissez-faire), we are each free to pursue our interests. We regulate our activity to achieve those interests because of competition, that is, others acting in their own interests in the same marketplace. Prices rise and fall based on availability and desirability of the products and services offered, and the community is served by the outcome moderated by competition. Adam Smith argues that under a laissez-faire system, competition in the marketplace would serve to regulate quality, price, and quantity.

In his view, it would also reward innovative and quality producers and punish gougers and charlatans. Smith based his views on assumptions of fair play and equal access to materials and to the marketplace.

By the mid-nineteenth century, manufacturers and other business enterprises were often incorporated. Applications for incorporation were subject to careful scrutiny and often approved subject to conditions or restrictions. In principle, applications for corporations were evaluated in terms of the public good, that corporate activity "should advance some specific public purpose."[20] The practice was often corrupt, however, and by the mid-1800s, movements were afoot to liberalize incorporation procedures. The belief in the principle of laissez-faire and the argument that people had a right to incorporate were powerful arguments in favor of changing public policy. In the view of reformers, corporations should not be subject to constraints of achieving some public good while in the pursuit of the shareholders' interests. For the most part, laws were changed, removing the barriers and the requirements for specific public benefit or purpose for incorporation.

In its earliest forms, the corporation sole was a means of investing institutions with rights constituted beyond those of an individual; the bishop as a corporation represents more than himself and acts with a broader moral authority, as it were, than other persons. These institutions, by definition, had a moral purpose or public good at their core. The corporate sole served the public at large to the extent that it enabled these institutions. Extending the corporate veil, that is, creating the corporate aggregate in the English system, to include groups such as aldermen and the like also serves this public benefit, at least in theory, by attracting good persons to public office and by limiting their personal liability. Even as corporations became more commercial, there was a tacit recognition that there was a public benefit to extending rights normally reserved to natural persons to groups of investors in a commercial enterprise. As Adam Smith would have it, "...by directing that industry in such a manner as its produce may be of the greatest value, he intends only his gain, and he is in this, as in many other cases, led by an invisible hand to promote an end which no part of his intention."[21]

The expectations of the corporation as a moral agent changed as the nature of the corporation changed. Later corporations of the industrial revolution saw little obligation to the public or to anyone beyond stockholders. In my view, the corporate awareness to community obligations may be coincidental with the shift from local to distant ownership; that is, as business ownership was no longer connected to any particular community, it ceased to recognize obligations to any such community. It has been argued, most notably by Milton Friedman,[22] that the only duty business has is to pursue a profit and that any effort to require a social benefit or mission to the business represents an unfair tax and burden. Friedman's view is repeated often, occasionally with enthusiasm, but it is not widely held in society at large.

If Adam Smith was correct, the marketplace would soon act against the interests of a tradesman whose quality was poor or whose product was cheaply made or dangerous. Perhaps, the invisible hand could be anticipated to act in the interests of a corporation that had a commitment to a community and acted as a good citizen. In self-interest, if not duty and obligation, corporations have acted as good citizens, at least insofar as such acts did not conflict with the economic interests of the corporations. Since the early 1990s, there has been growing recognition and awareness of the interests and rights of corporate stakeholders.[23] How is a business organization employing professional designers to balance its interest and the obligations of its employees? The moral minimum[24] of corporations has changed as society has changed. The earliest corporations aggregates were seen to have obligations limited to the actual business transaction; this was known as fair play.[25] Fair play itself was limited to the payment of debt to employees and suppliers, payment of taxes, and meeting the obligations of contracts. As corporate power and influence have grown, so has the "moral minimum" of corporations. In the late nineteenth and early twentieth centuries, the labor movement emerged in recognition of worker's rights and corporate obligations. In the second half of the 1900s, issues of product design and industrial pollution became important social concerns and corporations began to recognize obligations to consumers, to the environment, and even to the future. In all of these cases, the social awareness of the issues preceded the shift in the corporate point of view. The contentious period between social awareness of rights and corporate recognition of obligation could be seen as the period when the threshold of the moral minimum is raised.

The emergence of stakeholder theories of corporate governance has occurred as a reflection of the growth of corporate power and influence. Society has provided corporations with the special status under the law, but corporations have grown beyond the limits of a state or nation. The evolution of the multinational to global corporation has created wealthy and powerful organizations without local ties or loyalty. If such a corporation does not like the laws of one country, it has the capacity to move its operations to a place that it finds more suitable. International trade agreements may even supersede local or national law.[26] Under such circumstances, corporations may be seen as supernational, able to influence laws and treaties. "Globalization creates a borderless world, and at the same time dissolves the constraints on corporate conduct that operate within national boundaries."[27] To whom does the supernational corporation answer? Only its shareholders and its stakeholders. Clearly, the moral minimum must also change to reflect this new reality. The crux of the question of stakeholder theory in a global economy, where corporations exist in many nations and can influence even international law, is: why should they behave ethically?

If Friedman was correct and the only responsibility of business is to make a profit, then the entire corporate enterprise is morally neutral. In such a construct, corporations must obey the law and good commercial practice but no more; there is no moral minimum, only the threshold of law and practice. There is no stakeholder burden beyond what is required by often malleable legislatures. Friedman's view, however, requires individuals within the corporation to act morally neutral and to check their values at the company door, as it were. A view of the morally neutralized corporate employed is itself problematic. The question of the moral agency of corporations is debated less often in the global economy. There is broad, if not universal, recognition that corporations have a public as well as a private role.[28] It is suggested that broad obligations are as simple as enlightened self-interest, as is often suggested, but in such a view, the obligations of the corporation are dictated by the balance sheet and not by moral reasoning. The form of enlightenment is dismissed by Kant: "Men are honestly served; but this is not enough to make us believe that the tradesman has so acted from duty and from principles of honesty: his own advantage required it."[29] In fact, the emergence of stakeholder theory, in all of its permutations, is tacit recognition of the public duties and obligations of the corporation. Though many arguments are made, the argument for stakeholder theory often appeals to Kant, the Categorical Imperative and the concept of universalizability: "never to act otherwise so that I could also will that my maxim should become a universal law."[30] In the end, corporations, even multinational corporations, are composed of individuals and through the acts of individuals act with intentionality and are capable of moral reasoning.[31] As such, they have a duty to behave in such a manner as they would want the world to be organized and, in turn, be treated. In the end, in the global economy complete with multinational corporate persons (and personalities), we have little choice but to hold to such fundamentals. These issues are made more important when we consider professional obligations, environmental quality, and sustainability.

Business and Sustainability

It must be fair to say that most design efforts originate in commerce. Demand for design commonly begins in response to a real world need or opportunity and is commonly brought to designers by actors in the economy. For purposes of this discussion, we will consider "business" to include elements of government, since public entities also reach out to the private sector for design services, and in that way, governments are an important factor in the design economy. Governments also regulate the design community by establishing standards or policies that direct design and outcomes. While differences certainly do exist between public and private clients, they all share the role of being the originator of demand for design. Even if governments act as the reviewers or regulators of design, they are driving the character of demand and therefore the outcomes of design. Whether a client is a private or public concern does not change the ethical duty of the design professional and arguably the client.

As the source of demand, business drives design activities, but designers are constrained by things other than client demand. Designers have obligations to public welfare, for example, and to the profession and other professionals. They are obliged by the canons of the professions and by standards of practice. Of course, they also operate in a world of contracts, client expectations, and the demands of their own business interests. So, to a significant degree, the client is obliged by the obligations of the designer as well; that is, regardless of expectations, the client should expect the designer to act in a way that is not consistent with her professional ethics or obligations.

Business people, on the other hand, have no such canon to adhere to or to guide their behavior. To find the source of an ethical obligation, business has to look inward, to the values and ethics of its owners and managers. Some argue that the only obligation of business as an enterprise is to return a profit to its owners, but this is clearly short-sighted and incomplete. The limited nature of such an outlook suggests that the business exists and operates in a vacuum, that it does not have impacts beyond its immediate needs and desired outcomes. Clearly, the underlying facts of the sustainability conversation is the externalities wrought by such a narrow, poorly considered view. Adam Smith argues that the marketplace would react to an enterprise that wreaks havoc in its pursuit of profit alone, but in the modern economy, the externalities are often far afield or difficult to assess once in the marketplace. Firms may ship their problems, pollution, and waste to be out of sight or at least out of mind, to avoid the costs. The costs may include direct costs of waste but might also include indirect costs of community and environmental health, working conditions, and equity. As consumers become more aware of community impacts and workers confront unsatisfactory working conditions, demand for improvements may grow as well. Businesses may choose to improve performance and working conditions as a result of such pressure, but the response is self-interested and not a reflection of an ethical duty. The improved conditions may slip away by moving to a different area or country.

At a minimum, commerce is based on that trust between the actors involved that parties to a transaction are fundamentally honest. Contracts and agreements may outline the conditions and expectations of a transaction, and they may be necessary, on occasion, to compel one part or another to comply with those conditions, but there is an assumption that for the most part and in most circumstances, the parties will meet the commitments outlined in the agreement. Failure to do so would make future transactions less likely between the parties. Continued business success is based on reputation as well as performance, but this suggests that the standard of behavior is based primarily in self-interest for business, as opposed to an ethical obligation. In practice, businesses have relationship with employees and communities and even other businesses that require cooperation and perhaps even collaboration, but these could be seen as being done in the service of profit. Perhaps, an argument could be made that concern for the environment and sustainability is likewise in the business's interest; a desire for sustainable business outcomes and environment should be seen as part of doing business—the health of the community and nature directly impact the environment in which business operates. Environmental concerns and sustainability, however, require a longer view than provided for in most business plans and models. When measuring performance in quarterly or even annual reviews, the long time frames of environmental quality and sustainability may be lost.

When business people recognize that they have an obligation to the environment, to people, as well as to the goals of the business, then a true interest in sustainability will take root. Environmental stewardship and sustainability do require a long game, so to speak. The motivation for the long game in business requires a deeper and richer sense of the mission of the enterprise. As awareness of the breadth of impacts of the global economy has grown, some enlightened business leaders have embraced their role as actors in the effort to foster a sustainable world, and many have taken steps toward that goal. They adopt a practice of environmental and sustainability ethics based on the sure knowledge that an obligation to the future is good business and it is simply the right thing to do. In essence, they come to recognize that they do have an obligation to others as well as the owners of the business, and they manage and organize in such a way to include those principles in the structure, mission, and outcomes of the business. While it remains true that most businesses have not adopted such enlightened practices, the number of businesses attempting to do so increases from year to year.

Growth in Sustainable Business Practices

During the 1990s, a good many companies began to publish annual sustainability reports. The first environmental management standard appeared in 1992. Two years later, John Elkington began to promote the "triple bottom line," where "people, planet and profit" were all to be considered in how a business would conduct its affairs.[32] Where the traditional measures of business success were easily measured in terms of dollars, there is no such convenient measure for environmental and social concerns. While some aspects of the environment may be quantified as environmental "services," these are often limited to services to people or society and not to nature writ large. Also, while such efforts may have use, there are significant philosophical issues with measuring the environment only in terms of dollars. Likewise, social equity and welfare are challenging as well.

It may be necessary for organizations to have unique measures that do not directly equate to, from one company to another. Likewise, the measures might vary from one geographic location to another, from one business entity to another, or even at different stages as an enterprise grows and matures. If measures vary, how are they weighted against one another? And how are different companies compared with one another? It is the difficulty of bringing a linear method of measurement to a nonlinear system that facts are that appropriate measures will vary from organization to organization, from place to place, and even from time to time. The difficulty in finding specific methods that do relate to one another represents what is likely to be an ongoing effort for some time, perhaps an effort that requires constant upgrading and adaptation as the circumstances for a company change. In the end, the best methods may remain open to debate and question as to whether they have "value" to shareholders, the environment, and society. Perhaps, this is a good thing, since it will require diligence and constant attention to the issues of sustainability.

A review of some of the literature on the topic is revealing. The need for quantifiable measures is absolute, but when reading through the list of possible measures, one is struck by the absence of qualitative measures as well. There seems to be an assumption that if everything else is measured and the measures indicate that things are "good," then health, happiness, and all that is good will result. It reveals one aspect of measuring the triple bottom line—values and issues are different everywhere you go and even within the community and in different locations of the same company. Thankfully, diving deeply into those considerations is beyond the scope of this effort, but it is important to acknowledge.

Still, business plays a key role in the design of the world and the future. An 8-year study completed by the Massachusetts Institute of Technology Business Review and the Boston Consulting Group found that while 90% of business executives acknowledged that sustainability was important, only 60% of their organizations had any articulated sustainability strategy.[33] The study found that those firms with the greatest success in achieving sustainability had clear engagement from corporate leadership and had developed goals and a sound strategy to achieve the goals, including changing the business model if necessary. The study acknowledged that such business leadership is still uncommon but that those companies often find that the effort toward sustainability brings innovation and efficiencies that make contributions to corporate value. Indeed, many of the more forward-looking business executives have found sufficient value in increased efficiency, waste reduction, and the discovery of management opportunity to sufficiently justify the effort by using only the traditional measures of success.[34]

Most businesses are still not engaged in serious efforts to improve their impacts, but progress is being made. There is a good amount of "green wash" out there. Sustainability has become a buzz word with good reason. Research demonstrates that consumers show a definite preference for products and services offered by businesses that express the values of sustainability; 66% indicated a willingness to pay more for products and services sold by "sustainable brands."[35] In 2015, global consumer goods offered by sustainable branded companies grew by 4% compared with only 1% for companies not so identified. Clearly, there is an incentive to represent a business as sustainable, whether the claim is accurate or not. It is heartening to see the business expenditures on sustainable products and services that would be necessary if the commitment is more than a mere marketing claim; it is expected that, by 2019, business will spend more than a billion dollars on such services.[36]

The difficulty for the marketplace will always be to know how to tell the difference between the company committed to sustainability and the company with a marketing strategy to brand itself as sustainable. There has been no shortage of business organizations that have emerged to tout the bona fides of its green membership—sorting through them is another matter entirely. Few offer clear and transparent measures that would allow an assessment of the claims. It is in the measures, the key performance indicators (KPIs) of the company, where the tale may be told.

The KPIs are established by corporate leadership and reflect the commitment of the leadership to sustainability. What is important in the KPIs is the scope—not only what is being measured but also the progress shown over time. How KPIs are measured is challenging too. The adage that "what gets measured gets done" is particularly relevant. There are many number of sound approaches to the issue of measuring sustainability. As one might suspect, many are industry-, regional-, or even company-specific. The Global Reporting Initiative (GRI) offers one fairly robust and detailed method for organizations to measure and report progress toward sustainable operations and serves as an example for this discussion.[37] It describes the breadth and depth of comprehensive reporting of sustainability efforts for an organization.

The GRI standards are both broad and deep and are intended to be adapted to a specific application rather than taken as whole cloth and draped over an organization. The GRI represents more of a process than a standard. To employ the GRI, an organization must complete a close assessment of itself and its relationships in the world. To do this requires resources, expertise, and, above all, leadership. Sustainability begins at the top.

Leaders communicate the values and mission of an organization. Sustainability is more than a mere branding exercise and requires an organization to reconsider itself from the ground up, which means it must start from the top down. Once again, there is not a single best way to do this, but successful and easily communicated models exist. One example of such a model is The Natural Step. The Natural Step was developed by Sweden's Karl-Henrik Roberts in 1989 and is based on sound science and business principles. Today, The Natural Step is both a model and an organization. In brief, the model sets forth guiding principles:

"In a sustainable society, nature is not subject to systematically increasing…

1. Concentrations of substances from the earth's crust (such as fossil CO_2, heavy metals and minerals).
2. Concentrations of substances produced by society (such as antibiotics and endocrine disruptors).
3. Degradation by physical means (such as deforestation and draining of groundwater tables).
4. And in that society there are no structural obstacles to people's health, influence, competence, impartiality and meaning."[38]

Of course, moving from principle to practice requires a good deal of effort and expertise, but principles can serve as the means of communicating commitment, values, and mission in a fairly accessible form. For leaders and organizations committed to sustainability and the future, the way forward is made possible by models such as The Natural Step and processes such as GRI. In the final analysis, though, the commitment must first be understood and made in the boardroom. If designers are to conceive and design a sustainable future, the role of business and government as buyers of those services cannot be understated.

Why Should Corporations Be Viewed as Having Moral Responsibility?

The nature of the question seems to suppose that there is a basis for excepting corporations from having moral responsibility. In fact, there is no logical basis for the assumption. The most common reasons given for excusing corporations from moral responsibility are that because of the singular purpose of corporate existence and the fact that there is no single moral agent, corporations lack the necessary intentionality to be morally culpable.

Corporations were created and protected by law originally because it was perceived that there was some public benefit from such associations. Although I could find no reference to corporations in Adam Smith's *Wealth of Nations*, it is not difficult to sense that Smith would extend his theory of the "the invisible hand" to include the beneficial effects of a corporation seeking its own best interest. In a simple model, corporations once created under the law are able to use the social infrastructure of the courts, contracts, labor, capital, and markets to seek their best advantage. In doing so, wages are paid, goods and services are exchanged and distributed, and, all else being equal and above board, stockholders and society are better off. More than simply an organization to facilitate business, the corporation is recognized as a person under the law insofar as contracts and certain rights are concerned. Indeed, corporations are allowed rights and privileges unavailable to citizens. For example, in the United States, corporations will be free to enjoy certain rights under the Constitution and are free from having to extend those rights to employees in the workplace.[39] In essence, the allowance under the law to incorporate allows the creation of at least a limited person. Clearly, the state perceived that some public benefit would be achieved through such an arrangement.

The argument against corporate moral responsibility in general is based on the premise that corporations are amalgams of people and that decisions within the corporation are not moral because no one person has the agency, as it were, to form moral reasoning on a corporate scale. There appears to be an assumption that ethical responsibility or moral reasoning can occur only at the level of a single human being. In fact, there is not any logical limitation of moral reasoning to a human person, and some humans are not capable of moral reasoning.[40] On the other hand, there is no logic or compelling basis for limiting the possibility of moral reasoning and responsibility to a person. If, for example, it could be demonstrated by some means that chimpanzees or dolphins were capable of the reasoning required for moral judgment, it would be logical to extend such consideration to their actions.

Corporations act with intentionality; that is, they identify actions and behaviors they view as beneficial or conducive to their interests and they set about to act in such a way to facilitate the desired outcome. Presumably, these outcomes are determined rationally, that is, by understanding the implications of the action. The law provides them the autonomy to pursue these outcomes. In this way, the corporation acts with the intent to achieve or bring about some rationally determined outcome. While it is true that corporate decisions are a process usually not in a single person's scope to act, the reasoned basis, the autonomy to act, and the intentionality of such rational acts comprise sufficient basis to be capable of moral judgment.

Corporate behavior, despite many decision makers, cannot be seen as accidental or irrational. The hallmark of good and successful businesses is the intentionality and predictability of their behavior. It is not rational to accept that some behaviors and decisions are reasoned and intentional but at the same time exclude them from moral responsibility. It is the capacity for reasoning, understanding, and intentionality that is the basis for moral behavior. Corporations should be viewed as having moral responsibility, simply because, if the organization is capable of reasoned and intentional actions in business decisions, it follows that the capacity exists for moral judgment and reasoning.

The Duty of the Corporation to Professional Employees

The question of the obligations for managers of professionals needs some attention. Does an employer or manager who supervises professionals have a responsibility to give deference to the professional duties and obligations of such employees? There is an assumption that the organization values or at least recognizes the standard of care and obligations of its professionals. Organizations that employ professionals do so for several reasons:

1. The professional provides services directly to the organization (organization is client).
2. The professional contributes to the product sold by the organization (professional produces or contributes to the production of something that is sold or distributed).
3. The professional's services are sold to clients directly (the organization provides professional services).

In each case, the organization benefits from the presence of professionals on staff rather than from purchasing the services from an outside, independent source. The organization perceives an advantage to having the professional on staff. An organization that hires a professional seeks to add the benefit of the specialized training and expertise that a qualified professional would be expected to have. Among the ways an organization would assess whether an individual was qualified would be to confirm the training and experience in the individual and to be sure that the candidate for hire held the appropriate credential or license and was a qualified member of the appropriate professional organization. The essence of a profession is not limited to the unique technical knowledge in itself but the character of the body of knowledge and the moral obligation to service embodied in professional ethics. The nature of the distinction of professional responsibilities is further underscored by virtue of the value of the service in the marketplace, the exclusivity of the profession itself, and their own methods of evaluating a qualified candidate. This tacit recognition is presumptive acknowledgment of the professional's obligations and standard of care.

Having recognized and acknowledged the professional's responsibilities, the employer cannot reasonably expect the individual to separate the special knowledge gained through the professional training from the moral obligations and standard of care. As noted earlier, Goodpaster calls this the Nemo Dat Principles (NDP) after a Latin proverb *nemo dat quod non habet* or "nobody gives what he doesn't have." Goodpaster restates it as "No one can expect of an agent behavior that is ethically less responsible than he would expect of himself. I cannot (ethically) hire to have done on my behalf something that I would not (ethically) do myself."[41] Although Goodpaster applied the principle to stockholders' expectation of managers, it is directly applicable to the relationship between manager and professional employee.

The professional has a duty to prevent harm that may go beyond the interest of his employer and may run counter to such interests. Florman has argued that professionals (engineers specifically) have no responsibility for determining good or bad. They assert that engineers, for example, should not have the right to choose for whom they work, arguing that, in itself, engineering or the acts of a professional are neither moral or immoral.[42] If they are to be worthy of being called professional, the engineer, architect. or designer must be subject to a duty or obligations greater than other citizens; by virtue of the special knowledge of their profession, they acquire the burden to make ethical judgments and be subject to the public's interests.

Unger undoes Florman's view that engineers should not work only on projects that they approve of and that clients have a right to engineering services. Unger disagrees with Florman and points out the American Medical Association's Principles of Medical Ethics, which allow physicians the freedom (except in an emergency) to choose who to associate with and whom to serve. Attorneys are also free to choose to whom they provide service. Why should engineers be different? Florman argues that if engineers were to select what they worked on and for whom they worked, it would be tantamount to giving engineers the right to decide the form and format of society.

Unger also presents engineers' obligations and responsibilities as the concept of informed consent. He writes that it can be expressed in three parts:

1. The right of every individual potentially affected by a project to participate to an appropriate degree in the making of decisions concerning the project.
2. Engineers must do everything in their power to provide complete, accurate, and understandable information to all potentially affected parties.
3. Engineers should carry out, to the best of their ability, assignments that have been appropriately approved by all parties, even if they believe that they may have harmful consequences.

This third element exists to take away the power of the engineer to have, in effect, a veto power over the decision reached by stakeholders. This presents a problem with the engineer's duty to protect the public health and safety and the protection of the interests of the future. As Unger observes, "As a professional, an engineer is not an ordinary employee and hence cannot simply carry out assignments without giving some thought to the broader consequences."[43]

The question is then raised whether managers must defer to the professionals' opinion. In the end, the answer must be no. Managers have a responsibility to listen to the professional and then to act in a manner that is ethical. Managers may have knowledge that the professional does not have or may be working with conflicting recommendations from several professionals. The manager's burden to make a decision must ultimately be guided by facts, pertinent other information, and the ethical values of the organization and the individual. The manager has an obligation to make the best decision in light of the information at hand. Recommendations or concerns from staff professionals may be only one, albeit an important, source of information, and in the final analysis, every ethical decision is a personal one. Still, this raises the issue that if the professional is aware of an act by others deemed harmful or unethical by his professional standard of care, does he have a duty to act, to protect?

Professionals find themselves in one of two employment circumstances: working for another professional or working for a nonprofessional. In essence, the supervision is in the hands of someone who presumably has the same special knowledge and works under the same professional ethical obligations. Or, the professional is supervised by someone who has demonstrated that he or she does not share the same minimum level of competence and does not necessarily have the same professional ethical obligations. In the latter case, the supervisor has no reason to fear sanction by peer professional or the peak group. In this circumstance, the professional is the moral agent in charge of her decisions; if she does not have the autonomy to make such judgments, she is no longer able to act professionally or serve the interests of the firm, client, or stakeholders in a professional context. The professional being supervised by another professional retains the burden of assessing and acting as an autonomous agent, even though there is a supervising professional, but he or she may appropriately defer to the seniority and experience of a supervisor.

Design professionals are employed to provide access to their expertise by the employer and by extension to the employer's customers or clients. The distinction between customer and client is an important one. Clients are distinguished from customers in that the relationship between the professional and client is unequal. The clients are purchasing more than access to the professional's knowledge; rather, they are relying on the professional to act in their interest throughout the relationship. The word client is rooted in the Latin word "cliens," which describes a follower or a dependent. A customer exchanges value for value, purchasing goods or services but without any expectation that the seller will act in the interest of the buyer beyond the fair exchange of value; that is, "caveat emptor." Does the client have a relationship with a firm or individual professionals? Is the firm capable of meeting the obligations required in the relationship, or is this done by individuals within the firm? In truth, it might be difficult to draw a clear distinction, but in the end, are not the employees of the firm extensions of the professional in charge? If employees act under the direction of the professional in charge, then the professional is responsible for their work, as if he did it himself. There is no liability protection, nor is there an ethical caveat for acts made by others while under the direction of the professional. As noted in Chapter 3, several professional organizations require their members to report acts that may endanger public health and safety and the environment. Even though the professional employee may not be the professional in charge, he or she would retain the obligation to act on his or her knowledge of an unethical act.

The duties of those supervising professionals are of particular concern. Firms that hire professionals as employees have an obligation to acknowledge the standard of care and ethical obligations of the individual. It is the duty of the individual professional to assert these duties, should the circumstance arise. The duty of the professional remains an individual responsibility regardless of the circumstances; however, the supervisor bears the responsibility to honor these obligations. When a professional is hired as an employee, the employer assumes access to the special knowledge of the individual within the conditions of the standard of care and professional canon. In these circumstances, the employee has an obligation to act in the interest of the employer but always within the same obligations that would guide professional behavior with any client. Employers may reasonably expect nothing less. If the professional does not have the autonomy to make the necessary decisions, he is unable to fulfil his obligations.

The individual design professional has demonstrated the minimal level of competence acceptable to the profession. The profession, through a peak organization or other body, determines what the minimal level of competence consists of, describes the nature of conduct expected, and, when required, may make determinations on the appropriate standard of care. It is the individual designer, however, who acts in the marketplace to synthesize solutions to design problems, who promotes the profession through accomplishment, and who bears the responsibility for his work. The profession constructs the framework in which the individual acts, but ultimately, it is the individual who must act. This is true, whether the individual acts alone as an autonomous designer or within a firm.

Designers typically are employed in organizations and do have obligations to employers, the public, and clients. How can professionals balance these obligations when they are in conflict? The character of a profession is such that the practicing professional has specialized knowledge or skills that are acquired from education and practice. A profession is further distinguished from other work in that the knowledge is derived from a body of theory and knowledge that is dynamic in that it continues to grow and be refined. Further, the individual practitioner conducts his professional activities under the auspices of a larger professional community that serves as the arbiter of professional standards and conduct.

This principle of a standard of care does not lend itself to accurate description or specification. Instead, it is framed in statements like "A professional who is especially qualified in a particular field is expected to perform according to a standard of care that would be expected of any other similar professional under the same circumstances."[44] The standard of care is a critical element of professionalism. Society affords the various professions both status and privilege, but in exchange, it requires that they safeguard and act in the interest of the public. Clients rely on the professional for counsel, advice, and confidentiality. In the absence of acting in the public's interest, the professions face sanction. In the absence of client confidentiality and effective counsel, they have no object for their skill and no source of income.

In defining a profession, Firmage lists various attributes called for by "The Engineers Council for Professional Development (ECPD)...which has given the attributes of a profession as follows:

1. It must satisfy an indispensable and beneficial social need.
2. Its work must require the exercise of discretion and judgment and not be subject to standardization.
3. It is a type of activity conducted upon a high intellectual plane.
    a. Its knowledge and skills are not common possessions of the general public, they are the results of tested research and experience and are acquired through a special discipline of education and practice.
    b. Engineering requires a body of distinctive knowledge (science) and art (skill).

4. It must have a group consciousness for the promotion of technical knowledge and professional ideals and for rendering social services.
5. It should have legal status and must require well-formulated standards of admission.

ECPD has also stated what one who claims to practice a profession must do:

1. They must have a service motive, sharing their advances in knowledge, guarding their professional integrity and ideals, and rendering gratuitous public service in addition to that engaged by clients.
2. They must recognize their obligations to society and to other practitioners by living up to established and accepted codes of conduct.
3. They must assume relations of confidence and accept individual responsibility.
4. They should be members of professional groups and they should carry their part of the responsibility of advancing professional knowledge, ideals, and practice."[45]

Key among these is the requirement that the professional have a service motive at its core and that practitioners assume personal and professional responsibility for their acts. Ernest Greenwood writes that society allows a profession latitude to assume control over admission to the profession and the training of new practitioners but requires a fit between the professional culture and the norms of the society in which it operates.[46] The model of the professional is based on the traditional circumstance of the physician working as an autonomous professional.

Today, however, many, if not most, professionals work as employees within organizations and not as autonomous and independent practitioners. Where this used to be true primarily of design professionals, it has become commonplace among physicians and is increasingly more common among lawyers. As an employee, it is expected that the individual also has obligations to the interest of the employer. Presumably, the interests of all parties are not always in agreement; how can the professional balance these competing obligations when they are in conflict? When professionals are employees, to whom do they owe their professional fidelity and to whom are they duty bound? John Ladd observes that "It is obviously difficult to be ethical when one is powerless."[47] It would seem that in the absence of these important attributes, professions are at considerable risk of ceasing to meet the test of a profession. If technical knowledge and counsel are subject to bureaucratic interests and direction without recourse, then the public service foundation of the profession is no longer paramount.

However, by nature, "All professionals are moral enterprises that involve concerns beyond the application of technical principles. How well the professionals meet these moral obligations will determine the freedom of the individual professional enterprise."[48] This moral responsibility ascribed to professionals is forward looking. "It is about what people ought to do to bring about or prevent future states of affairs. It is based on the duty each one of us has to watch out for what may happen to others or to oneself. As such, it implies concern, care and foresight."[49] It would seem that the professional as an individual is responsible for making a judgment call based on the knowledge available to him. Ladd argues that professional ethics must be "basically an open-ended, reflective and critical intellectual activity.," that is "ethics must, by its very nature, be self directed rather than other directed."[50] His point is compelling. Professionals set themselves apart from others and believe that only another professional is able to assess the behavior of a colleague. Ladd asserts that if this is true, there are responsibilities that stem directly from the power of being a professional. Both individual and collective powers of professions require responsibility for the acts of the individual practitioner.

Alpern writes that "though engineers are bound by no special moral obligations, ordinary moral principles as they apply in the engineer's circumstances stipulate that they nonetheless be ready to make greater personal sacrifices than can normally be demanded of other individuals."[51] His remarks have a broad application beyond engineers. He makes his argument first by describing what he calls the Principle of Care: "Other things being equal, one should exercise due care to avoid contributing to significantly harming others."[52] This includes knowing what harm might result from our actions and taking steps to mitigate or avoid that harm. It refers to one "contributing" to harm, so the obligation extends to taking part in activities that might result in harm.

Alpern acknowledges that the Principle of Care is vague. The degree of care required is not always the same. He says, "the degree of care due is a function of the magnitudes of the harm threatened and the centrality of one's role in the production of that harm." He introduces the Corollary of Proportionate Care: "When one is in a position to contribute to greater harm or when one is in a position to play a more critical part in producing harm than is another person, one must exercise greater care to avoid doing so."[53] In light of the Principle of Care and the Corollary of Proportionate Care, and the role of professionals, they "can be held to a higher standard of care; that is, it can be demanded that they be willing to make greater sacrifices than others for the sake of public welfare."[54]

Does this require the professional to make unreasonable sacrifices? McFarland suggests the Kew Gardens principles as a starting point for assessing the moral duty of the professional.[55] The Kew Gardens principles involve four preconditions:

1. A critical need (a right or some good is threatened or at risk)
2. Proximity, in the area or in the case where responsibility is a function of notice or awareness

3. Ability to help, constrained only but not risking damage to oneself or duties owed to others
4. Absence of other source of help

The Principle of Care and Kew Gardens principles provide guidelines for determining whether the individual professional has an obligation to act. The professional is to be held to a higher standard of care than those without expertise or specific awareness, but there must be limits to what is expected. The Corollary of Proportionate Care provides the means to assess whether the individual must act. Still, the obligation cannot be limitless.

Justification for a Whistle-blower

What is the professional to do if he finds himself in untenable ethical circumstances? In her essay "Whistleblowing and Professional Responsibility," Sissela Bok discusses the conflict between organizational and professional obligations and the inherent difficulties that may confront individuals.[56] She correctly points out that, while codes of ethics routinely stress the professional obligations to public welfare, they also require loyalty to the profession.[57] In addition, though she doesn't mention it specifically, some professional codes of ethics include loyalty to the professional's employer.[58] Bok does report that some codes of ethics also require the individual to report or speak out in circumstances when health and safety are threatened. Bok notes that what is not addressed or provided is a guide for individuals to balance these sometimes-competing obligations. How then is the working professional to justify a decision to reveal information with which she has been entrusted? What justifies a professional in becoming a whistle-blower?

Bok describes how those who step outside an organization or profession to report some threat or risk do so at the risk of a high cost. Still, society relies to a degree on individuals to do this and benefits from their sacrifice. Bok reviews the elements or stages of whistle-blowing as being dissent, breach of loyalty, and accusation. She notes that "The more repressive the authority they challenge, the greater the personal risk they take in speaking out."[59] The nature of these elements is to threaten the organization and to invite various degrees of retaliation. Duska argues that there is no obligation of loyalty between an employer and an employee, and so, there is no breach associated with whistle-blowers.[60] In any case, Duska's point, regardless of how well made, is beyond further consideration in this paper. The expectations of employee loyalty are fairly deep in society, and it could be argued that professional employees have a duty equivalent to that of a trustee by virtue of the professional codes of ethics.

There is a parallel expectation on the part of the public that the professional will act in its best interests. This expectation is acknowledged in every code of ethics. This expectation creates a moral duty for the professional and the moral basis that supports whistle-blowing. The professions are not entitled to a paternalistic approach to this duty; the public or effected individuals have the right of informed consent. The idea that individuals should have the opportunity to decide for themselves on issues that affect them is a fundamental element of the right of self-determination. For a professional to decide to withhold information is, in fact, a decision to decide for others, to deny a person the right of informed consent, and to limit a person's freedom.

This, however, does not make the whistle-blower's task easier or less difficult. To be effective, the whistle-blower's information must be timely, specific, and of legitimate public interest. Whistle-blowing that occurs well after the fact of effect is generally not considered timely; likewise, broad statements of vague doom or impacts are not compelling. The issue must be fair and in legitimate interest of the public, not malicious or personal in nature. Given that one's information is timely, specific, and discloses a legitimate issue, how is the individual able to assess her responsibility? Bok suggests that the best chance the individual has is to seek objective counsel before she acts and to have the necessary facts to support the claim. This is thin promise to the individual, however, and highlights Bok's original concern that there is no formal guideline or basis for individuals to assess their obligation. This seems to be an important omission in light of the duty to report violations or threats to the public safety canon. Indeed, the American Society of Civil Engineers (ASCE) has the most comprehensive duty to report[61] of all of the codes of ethics reviewed and publishes a handbook *ASCE Ethics, Edict, Enforcement and Education*. The scope of the code of ethics and handbook is to describe the duty to report and the process by which ASCE will judge those charged with violations of the code, but it provides no guidance for the person making the charge.

Norman Bowie outlines the moral justification of whistle-blowing as requiring an appropriate motive and the exhaustion of all internal channels for the resolution of dissent. The whistle-blower must have compelling evidence to support the claim, and in Bowie's view, there must be some chance of success. He states that "In general, whistle blowing that stands no chance of success is less justified than that with some chance of success."[62] Michael Davis reports that the whistle-blowing can be either morally permissible or morally required, depending on the circumstance. Davis reports that the standard theory is that whistle-blowing is morally permissible when[63]:

1. An act, product, or service will "do serious harm to the public"
2. The whistle-blower is justified in believing that, after internal attempts to address the concern, nothing will be done to prevent harm
3. Internal channels and attempts to address the concern are exhausted,

Whistle-blowing is morally required when:

1. The whistle-blower has evidence that would convince a "reasonable, impartial observer"
2. The whistle-blower has good reason to believe that revealing the threat will prevent harm

Davis is concerned, however, that the standard theory does not account for the reality of whistle-blowers. He recounts three paradoxes:

1. Whistle-blowers act with considerable risk to their career, financial security, and relationships with others.
2. Harm is relative. Harm is supposed to mean physical harm and usually does not "stretch" to cover things such as injustice, deception, and waste.
3. Generally, whistle-blowers are unsuccessful.

Davis proposes another view that he calls the complicity theory of whistle-blowing. He observes that few whistle-blowers are simple third parties; instead, they are part of the organization. As part of the organization, the whistle-blower is morally complicit in the act of harm. This complicity theory is particularly applicable to professionals. According to Davis, a person is morally required to reveal information under the following circumstances:

1. What you will reveal derives from your work for an organization.
2. You are a voluntary member of that organization.
3. You believe that the organization, though legitimate, is engaged in serious moral wrongdoing.
4. You believe your work for that organization will contribute (more or less directly) to the wrong if (but not only if) you do not publicly reveal what you know.
5. You are justified in beliefs 3 and 4.
6. Beliefs 3 and 4 are true[64].

Davis's theory is an improvement over the standard theory and Bowie's view in several ways. First, the whistle-blower comes by the information honestly as an employee serving voluntarily in the organization. The complicity theory requires only that a serious moral wrong is in question and not the somewhat-more-ambiguous concept of harm, but the moral wrong must be serious so that not every problem is a cause for whistle-blowing. The whistle-blowers do not have to believe that their work contributes to the wrong or that it will be prevented if they act. Complicity theory does,

The Design Professional and Organizations 99

however, require proof. Complicity theory would seem to provide a strong model for the justification of whistle-blowing by a professional.

Having a moral justification for whistle-blowing, however, does not speak to the practical implications. A more practically oriented and specific set of guidelines was prepared by the Institute of Engineering Ethics Education (IEEE) for engineers.[65] These guidelines include the following:

1. Establish a clear technical foundation for the claim. IEEE encourages engineers to seek technical council from others and to listen to other points of view but cautions that it is not necessary to prove that a failure or harm will occur. It is only necessary that there is a likelihood or harm. Certainty is not required.
2. Arguments should be objective and on a high professional plane. Dissenters are cautioned to present their concerns as objectively as possible, appealing to the technical issues and leaving unrelated issues aside. Unrelated grievances and issues only serve to muddle the key points and discount the importance of the concern.
3. Concerns should be addressed as early as possible and at the lowest managerial level as possible. Dissenters are discouraged from waiting until a project has progressed to a point where a small problem becomes a large problem. Likewise, issues should be addressed according to the organizational chain of command beginning at the lowest level possible.
4. The professional is encouraged to be sure that the issue warrants this type of attention. If a problem or concern can be addressed through channels or through normal development processes, heightened treatment is not necessary. Also, the dissenter should be certain that the concern is significant and not simply a case of "standing too close" to the issue.
5. The concern should follow organization channels. The organization should be given every opportunity to respond to the concern. If normal channels are not successful, the dissenter should look for organizational dispute resolution mechanisms.
6. In all cases, the dissenter should keep records and collect paper. A detailed paper trail may become critical to make one's point and for self-defense.
7. IEEE recognizes that if a resolution is not possible, the individual may find it necessary to consider resigning from the position. Resigning tends to add credibility to one's claims but reduces the effectiveness of the dissenter within the organization. Once outside the organization, the dissenter no longer has access to data or the channels for discussion.

8. In some cases, anonymity may be considered. Contacting regulatory agencies or the press may be an option. Anonymous dissent, however, is often ineffective and easily deflected.
9. Outside resources may also be used to expose the issue. Going to an outside source such as a newspaper should be considered carefully. A person and an organization with known credibility and a reputation for fairness and professionalism should be found. The issues should be presented as before, dispassionately, focusing on the objective and technical characteristics and avoiding personal animus or grievances.

IEEE's purpose was a practical one: "The assumption here is that the engineer's objective is to prevent some serious harm, while minimizing career damage."[66] It cautions that the personal and professional costs of dissent and whistle-blowing can be daunting.

Professionals are held to a higher standard of care than other employees by virtue of their professional ethics and specialized knowledge. Justification for whistle-blowing can be found in several forms based on the models described by Bok, Bowie, and Davis, though Davis's complicity theory may be the most relevant for professionals. Elements common to all are the motivation for the act, exhaustion of alternatives, and the objectivity and relevance of the proof. The whistle-blower, however, is often not successful, and even if successful, he pays a very high professional and personal price for his action. A justification for whistle-blowing does not necessarily create an obligation for an employee to act. Individuals should consider the chances of success for their claim, a realistic estimate as to whether their act will lead to the desired outcome and whether the professional and personal cost is too great in light of the chances of success. In the end, however, the professional must recognize that some actions may be morally required, regardless of the professional or personal cost.

In the final analysis, professionals acquire responsibilities to their employers but not necessarily to the exclusion of their ethical obligations. Professionals in subordinate roles may defer to more senior, experienced professionals, but they should do so cautiously. It is a two-way street, however, and organizations owe consideration to professional employees and must recognize that the special skills for which they hired an individual come with obligations that may go beyond the more narrowly defined interests of the business or agency.

Endnotes

[1] McFarland, Michael C., "The public health, safety and welfare: An analysis of the social responsibilities of engineers," in *Ethical Issues in Engineering,* Johnson, Deborah, editor, Englewood Cliffs, NJ, Prentice-Hall, 1991, p. 159.

The Design Professional and Organizations 101

2. Based on the Kitty Genovese murder which occurred in the Kew Garden section of New York City, ibid, p. 63.
3. Donaldson, Thomas, "The ethics of conditionality in international debt," in *Ethical Theory and Business*, 5th edition, Tom L. Beauchamp and Norman E. Bowie, editors, Prentice Hall, Upper Saddle River, NJ, 1997, p. 646.
4. Pittman, *Strategic Management: A Stakeholder Approach* 1984 as quoted in "Business ethics and stakeholder analysis" Kenneth E. Goodpaster in *Ethical Theory and Business*, 5th edition, Tom L. Beauchamp and Norman E. Bowie, editors, Prentice Hall, Upper Saddle River, NJ, 1997, p. 76.
5. Freeman, R. Edward, "A stakeholder theory of the modern corporation," in *Ethical Theory and Business*, 5th edition, Tom L. Beauchamp and Norman E. Bowie, editors, Prentice Hall, Upper Saddle River, NJ, 1997, p. 66.
6. Goodpaster, Kenneth E, "Business ethics and stakeholder analysis," in *Ethical Theory and Business*, 5th edition, Tom L. Beauchamp and Norman E. Bowie, editors, Prentice Hall, Upper Saddle River, NJ, 1997.
7. Luban, David, Alan Strudler, and David Wasserman, "Moral responsibility in the age of bureaucracy," in *Moral Issues in Business*, 8th edition, by William Shaw and Vincent Barry, Wadsworth Thomson Learning, 2001, p. 47. Luban et al. describe the person working inside a corporation as having "fragmented knowledge" and use the lesson of Milgram's experiments to illustrate what they call the "The Psychology of Destructive Obedience." They assert that the less people know about the consequences of their actions, the less likely they are to be concerned with the outcomes. Person in organizations are seen as operating only with the knowledge necessary to perform their tasks and only with the authority necessary to complete the task. Tasks are completed in a sort of organizational-informational isolation. The task itself is merely a step or even a substep of a larger process of which the person is unaware or unable to evaluate.
8. ibid., To be able to make a moral choice four conditions must be met. The four knowledge conditions are:

   1. recognition of the "fork in the road," that a decision must be made
   2. awareness that the decision must be made in short order
   3. knowledge of the limited number of choices
   4. adequate information of knowledge necessary to make an informed choice

9. For purposes of this discussion "system" refers to the flows of information and materials that occur within an organization either formal or informal. These flows occur between person or groups of persons. Information or materials are sent from one person to the next so that the person will act. The objectives of the organization are served and rely on the timely completion of these acts.
10. Jackall, Robert, "Moral mazes: Bureaucracy and managerial work," in Donaldson, Thomas and Patricia H. Werhane, editors, *Ethical Issues in Business*, 6th edition, Prentice Hall, Upper Saddle River, NJ, 1999, p. 96.
11. Barcalow, Emmett, *Moral Philosphy Theories & Issues*, 2nd edition, Wadsworth Publishing, Belmont, CA, 1998, p. 2.
12. Business organizations and business people are perceived as ruthless and unethical. A survey of the general public found that 68% of those surveyed believed the behavior of executives was the primary cause of decline in business standards

and productivity (New York Times/CBS News poll, 1985). There are certainly examples of such behavior and often there are obvious rewards for such behavior. Try as we might it is often difficult to demonstrate that the unethical person is harmed for his behavior. It is easy to understand the motivation for such behavior: Self-interest, profit and winning. Many business people have argued that business is a unique undertaking and that it occurs outside of the bounds of normal society and cannot be held to the same measure as normal social behavior. Business is described as game with its own rules. In fact there is no bright line separating business from our personal lives; business activities are subject to the evaluation of society. The fact remains however that there are strong reasons for ethical practices in business (Shaw). Much of business is done on trust. Trust implies faith and confidence that another will meet their commitments and make good on promises. Is it possible for a person to lie and gain an advantage? Yes, but usually only once in a given relationship. Businesses and individuals can survive by misleading customers/clients but they rely on constantly having to find new customers. Without trust and being able to rely on a person's promise, no relationships are possible. Business ethics are formal and informal. Employees look to organization leaders as ethical barometers, not only what is tolerated but also what is expected. More important than any statement or policy are the behavior of supervisors and managers. The perception among workers of unethical behavior of supervisors and executives was often used as justification for poor behavior on the part of workers. Absenteeism, petty theft, poor performance on the job were all sited as acceptable behavior based on what workers observed in organizational leadership. Among employees surveyed, the concern for ethics on the part of the organization is an important reason to remain with an organization. The survey did not find any statistical difference between employees views of ethic in either the private or not for profit sectors (KPMG).

It is interesting to note however that the public may be willing to forgive an organization for unethical behavior but not individuals. Individuals judged to have been unethical often find the cost to be quite high: A loss of career, income and reputation. Organizations have demonstrated over and over again that image can be repaired, often by sacrificing an employee or two. Recent examples include the financial scandal in the United Way, the infamous Ford Pinto or the most recent Firestone tire problems (Beauchamp). Companies appear to increasingly value ethics on the part of employees. In a survey by KPMG 86.4% of the companies surveyed said they had some form of written statement about ethics and 72.7% reported having some kind of program or initiative that promotes ethics among its employees. Half of these have actual training programs to teach employees in a Code of Ethics and how to integrate ethics into day to day business practices. 88.5% of these companies included ethics issues in their risk assessment programs.

[13] Jackall, Robert, "Moral mazes: Bureaucracy and managerial work," in Donaldson, Thomas and Patricia H. Werhane, editors, *Ethical Issues in Business*, 6th edition, Prentice Hall, 1999.

[14] As defined in Black's Law Dictionary as "An artificial person or legal entity created by or under the authority of the laws of a state or nation, composed in some rare instances, of a single person and his successors, being incumbents of a particular office, but ordinarily consisting of an association of numerous

individuals. Such entity subsists as a body politic under a special denomination, which is regarded in law as having personality and existence distinct form its several member, and which is, by the same authority, vested with the capacity of continuous succession, irrespective of its membership, either in perpetuity or for a limited number of years, and of acting as a unit or single individual in matters relating to the common purpose of the association, within the scope of the powers and authorities conferred upon such bodies by law." Black, Henry Campbell, *Black's Law Dictionary*, 5th edition by the Publishers Editorial Staff, with contributions from Joseph R. Nolan, and M.J. Connolly, West Publishing, St. Paul, MN, 1979, p. 307.

[15] Walker, David M., *The Oxford Campanion to Law*, Clarendon Press, Oxford, 1980, p. 293.

[16] ibid, p. 684.

[17] Shaw, William H. and Vincent Barry, *Moral Issues in Business*, 8th edition, Wadsworth Publishing, Belmont, CA, 2000, p. 197.

[18] ibid, p. 198.

[19] Smith, Adam, *An Inquiry Into the Nature and Causes of the Wealth of Nations*, Oxford University Press, Oxford, England, 1993.

[20] Shaw, ibid, p. 199.

[21] Smith Adam, ibid. p. 292. It is important to note that Smith was writing about the behavior of an individual not that of a commercial corporation. In the Wealth of Nations Smith speaks only of corporations as civic organizations. Smith does write of trade or craft groups as corporations and the deliberate control or manipulation of markets by monopolies but does not anticipate the later forms of commercial corporations. Nonetheless, he believes monopolies must be policed and the market will then resolve such issues as price manipulation. (p. 61). The point is that Smith would appear willing to extend his view of individual behavior and the invisible hand to corporations.

[22] Friedman, Milton, "The social responsibility of business is to increase its profits," in *Ethical Theory and Business*, Tom L. Beauchamp and Norman E. Bowie, editors, Prentice-Hall, Upper Saddle River, NJ, 1997, pp. 56–61.

[23] Donaldson, Thomas, "The stakeholder revolution and the clarkson principles," in *Business Ethics Quarterly*, 12(2), 107–111.

[24] Simon, John G., Charles W. Powers, and Jon P. Gunnemann, "The responsibilities of corporations and their qwners," in *Ethical Theory and Business*, Tom L. Beauchamp and Norman E. Bowie, editors, Prentice-Hall, Upper Saddle River, NJ, 1997, p. 62.

[25] Fair play may have been further limited into fairness among one's own kind since fairness to people of other ethnicity or even nationality was often at best relative.

[26] The North American Free Agreement for example provides for vehicles from one member nation to travel over the highways or the other member nations. Vehicle safety standards and requirements in the United States are more stringent than in Mexico so trucks from Mexico have been barred from traveling into border states. Mexico has claimed a violation of the treaty and sought relief under the NAFTA agreement.

[27] Collier, Jane and John Roberts, "An ethic for corporate governance?" in *Business Ethics Quarterly*, 11(1), 68, 2001.

[28] Cragg, Wesley, "Business ethics and stakeholder theory," *Business Ethics Quarterly*, 12(2), 136, 2002.
[29] Kant, Immanuel, *Fundamental Principles of the Metaphysics of Morals*, first published in 1785 translated by T.K. Abbott, Prometheus Books, Amherst, NY, 1988, p. 22.
[30] ibid, p. 2.
[31] Russ.
[32] Elkington, John, *Cannibals With Forks: The Triple Bottom Line of 21st Century Business*, Capstone Publishing, Mankato, MN, 1999.
[33] David Kiron, Gregory Unruh, Nina Kruschwitz, Martin Reeves, Holger Rubel, and Alexander Meyer Zum Felde, "Corporate sustainablity at a crossroads—Progress toward an our common future in uncertain times", https://sloanreview.mit.edu/projects/corporate-sustainability-at-a-crossroads/, May 23, 2017, accessed March 26, 2018.
[34] "From reaction to purpose: The evolution of business action on sustainability" https://www.theguardian.com/innovative-sustainability/2017/oct/31/charting-the-course-of-sustainability-in-business-from-the-1960s-to-today, accessed March 26, 2018.
[35] University of California Riverside, Major Corporations' Growing Interest in Sustainable Product Design, https engineeringonline.ucr.edu/resources/infographic/major-corporations-growing-interest-in-sustainable-product-design/ 2015, accessed April 10, 2018.
[36] ibid.
[37] GRI, GRI Sustainability Reporting Guidelines, G3.1 Reference Sheet, https://www.globalreporting.org/resourcelibrary/G3.1-Quick-Reference-Sheet.pdf accessed May 6, 2018.
[38] The Natural Step, https://thenaturalstep.org/approach/ 2018, accessed May 20, 2018.
[39] Corporations have the right of free speech but in the workplace employees do not. Certain other rights are denied corporations however. For example corporations do not have constitutional protection from self-incrimination and may be made to testify against their own interest.
[40] Infants, sleeping, unconscious, drunk or insane people cannot form the necessary judgment to be considered moral agents. We might observe that an insanity defense at a criminal trial is a formal recognition that a defendant is not a moral agent.
[41] Goodpaster, Kennth, "Business ethics and stakeholder analysis," in *Ethical Theory and Business*, Thomas Beauchamp and Norman E. Bowie, editors, Prentice Hall, Upper Saddle River, NJ, 1997, p. 83.
[42] Florman, Samuel, *The Civilized Engineer*, St. Martin's Press, New York, 1987.
[43] "Codes of Engineering Ethics," Stephen Unger in Johnson, Deborah, editor, *Ethical Issues in Engineering*, Englewood Cliffs, NJ, Prentice-Hall, 1991, p. 112.
[44] Firmage, Allan D., "The definition of a profession," in Johnson, Deborah, editor, *Ethical Issues in Engineering*, Prentice-Hall, Englewood Cliffs, NJ, 1991, pp. 63–66.
[45] ibid, Firmage "The definition of a profession," D. Allan Firmage, pp. 63–66. Firmage reports on the findings of "The Engineers Council for Professional Development (ECPD)…which has given the attributes of a profession as follows:

1. It must satisfy an indispensable and beneficial social need.
2. It work must require the exercise of discretion and judgment and not be subject to standardization.
3. It is a type of activity conducted upon a high intellectual plane.
    a. Its knowledge and skills are not common possessions of the general public, they are the results of tested research and experience and are acquired through a special discipline of education and practice.
    b. Engineering requires a body of distinctive knowledge (science) and art (skill).
4. It must have a group consciousness for the promotion of technical knowledge and professional ideals and for rendering social services.
5. It should have legal status and must require well-formulated standards of admission.

ECPD has also stated what one who claims to practice a profession must do:

1. They must have a service motive, sharing their advances in knowledge, guarding their professional integrity and ideals, and rendering gratuitous public service in addition to that engaged by clients.
2. They must recognize their obligations to society and to other practitioners by living up to established and accepted codes of conduct.
3. They must assume relations of confidence and accept individual responsibility.
4. They should be members of professional groups and they should carry their part of the responsibility of advancing professional knowledge, ideals, and practice." pp 63–64.

[46] Greenwood, Ernest, "Attributes of a profession," in Johnson, Deborah, editor, *Ethical Issues in Engineering,* Prentice-Hall, Englewood Cliffs, NJ, 1991, pp. 67–77.
[47] Ladd, John, "Collective and individual moral responsibility in engineering: Some questions," in Johnson, Deborah, editor, *Ethical Issues in Engineering,* Prentice-Hall, Englewood Cliffs, NJ, 1991, pp. 26–39.
[48] ibid, p. 65.
[49] ibid, p. 36.
[50] Ladd, John, The quest for a code of professional ethics: An intellectual and moral conflict, in Johnson, Deborah, editor, *Ethical Issues in Engineering,* Prentice-Hall, Englewood Cliffs, NJ, 1991, ch. 3, sec. 12, 130–136.
[51] Alpern, Kenneth D., "Moral responsibility for engineers," in Johnson, Deborah, editor, *Ethical Issues in Engineering,* Prentice-Hall, Englewood Cliffs, NJ, p. 187.
[52] ibid, p. 188.
[53] ibid, p. 189.
[54] ibid, p. 189.
[55] McFarland, Michael C., "The public health, safety and welfare: An analysis of the social responsibilities of engineers" in Johnson, Deborah, editor, *Ethical Issues in Engineering,* Prentice-Hall, Englewood Cliffs, NJ, p. 159.
[56] Bok, Sissela, "Whistleblowing and professional responsibility," in *Ethical Theory and Business,* 5th edition, Tom L. Beauchamp and Norman E. Bowie, editors, Prentice-Hall, Upper Saddle River, NJ, 1997, p. 328.

57. Nearly every code of ethics includes a statement about the individual's duty to the profess beginning with the Hippocratic Oath "To hold my teacher in this art equal to my own parents; to make him a partner in my livelihood....:" as written in *Ethics, Tools and the Engineer* by Raymond Spier, CRC Press, New York, 2001, p. 214.
58. Engineering peak organizations often include the phrase "Shall act in professional matters for each employer or client as faithful agents or trustees" in the Canons of ethics and "Being honest and impartial and serving with fidelity the public, their employers and clients..." from American Society of Mechanical Engineers. Similar working is found in the National Society of Professional Engineers and the American Society of Civil Engineers (ASCE).
59. ibid. Bok, p. 329.
60. Duska, Ronald, "Whistleblowing and employee loayalty," in *Ethical Theory and Business*, 5th edition, Tom L. Beauchamp and Norman E. Bowie, editors, Prentice-Hall, Upper Saddle River NJ, 1997, p. 335.
61. the ASCE Guidelines to practice Under the Fundamental Canons of Ethics states in part-Canon 1.c "Engineers whose professional judgment is overruled under circumstances where the safety, health and welfare of the public are endangered, or the principles of sustainable development are ignored shall inform their clients or employers of the possible circumstance.

    Canon 1.d. Engineers who have knowledge or reason to believe that another person of firm may be in violation of any of the provisions of Canon1 shall present such information to the proper authority in writing and shall cooperate with the proper authority in furnishing such further information or assistance as may be required.
62. Eighth Edition, William H. Shaw and Vincent Barry, *Moral Issues in Business*, Wadsworth Publishing, New York, 2001, pp. 378–379.
63. Some Paradoxes of Whistleblowing, Michael Davis, in moral issues in business, 8th edition, William H. Shaw and Vincent Barry, Wadsworth Publishing, New York, 2001, p. 421.
64. ibid.
65. Guidelines for Engineers Dissenting on Ethical Groundsonlineethics.org
    The Online Ethics Center for Engineering and Science
    Guidelines for Engineers Dissenting on Ethical Grounds
    IEEE Ethics Committee 11/11/96.
66. ibid., p. 1.

5

The Choice for Sustainability

Being less bad is not the same as being good. It is only being less bad.

William McDonough

With the ethical obligation to produce sustainable outcomes come practical considerations. Design professionals work in an environment where their work is reviewed by others, in some cases others with far less knowledge of the work. Their work is evaluated within a framework of regulations, laws, past practices, desired outcomes, budgets, and competition. Every project is subject to the acceptance of the marketplace, concerns of insurers, and ultimately professional peers in terms of reflecting an accepted standard of care. In the end, even the most autonomous designer works for someone, and the result of the work is used by others. No professional designer works in a vacuum. The designer is expected to balance the concerns of the client, of the end user, and of the context of where the design will be used or built.

The designer's special knowledge is assumed to include awareness of legal or public standards applicable to the end product as well as how to select and use various materials for their best advantage in construction. The client expects the designer to understand the budget constraints and the expected outcomes of the project. Users made up of individuals and communities also have expectations and interests to be considered. Communities themselves may have diverse interests. The interests of the formal political and administrative bodies may differ from those individuals or groups of individuals that make up a community. The professional's own firm has a business interest in the work, ranging from its economic value to the firm to maintaining a long-term relationship with a favored client and its contribution to the firm's reputation. At a personal level, the designer's work becomes part of the lasting image of the designer in the marketplace and an important component of a professional reputation.

The designer must synthesize a solution that responds to all of these stakeholders. Concerns with sustainability add the interests of the environment and the well-being of people today and in the future. Well-being implies not only satisfaction but also a meaningful contribution to the health of the community and the individual. Balancing the interests of the various stakeholders is a political exercise. Each stakeholder promotes the attributes of the project that serve his or her own interests. Of course, not every stakeholder is equal in the process, and clients and regulatory agencies might be expected to have greater weight than, say, unorganized community groups or individuals. The designer's own firm and supervisor might have more influence than a knowledgeable interested colleague. In the end, it is the designer who must bring order to chaos, balance these diverse, often adversarial voices, and produce a design. Among all these voices, how are the interests of a flourishing environment and society to be heard?

The Design Professional as Leader

Design today is a complex process. Designers have a central leadership role in the process, but they are by no means the only participants. They are charged with finding a solution that meets the objectives of the client and users and balances the interests heard in the chorus of competing interests that we refer to as stakeholders. The designers lead this process. Florman argues that they are not even the principle players in the process, that the wishes of employers, clients, and public have more weight than the individual designer. As discussed earlier, however, the designer who has no authority in the creation of his work is not acting in professional capacity. The designer serves these stakeholders as the leader of the process, attentive and responsive, but leader nonetheless. If this was to be otherwise, the individual designer would not be in a position to act as a professional but would instead become a workman with special skills, like a highly qualified machinist or a cabinet maker taking orders from his stakeholders. Clients would become customers, not requiring or paying for access to the special knowledge and training of the professional but instead merely buying drafting services.

The designer is deemed professional through demonstration of special knowledge and incurs ethical and professional obligations in the conduct of his work. The standard of care is dynamic and always subject to change, as the professions adapt to the world in which they operate. The concern with a wider application of the standard of care to include diverse stakeholders has emerged over the last 50 years. This reflects a practical adjustment to the designer's scope of concern as others demonstrated and claimed interests in the outcomes of the designer's work. A span of stakeholders is now accepted for the most part as how business is done. Sustainable design adds interests to these considerations. However, the interests of sustainability are silent, often without a discrete representative among the other parties. In this way, sustainability is like or even part of the designer's interest in quality. There is a presumption that the minimal level of competence required to become a professional is a signal that the designer is qualified and the product of her work meets some undefined test of quality. The obligation of the professional to consider issues of sustainability amends the standard of care and changes the admittedly loose definition of quality. With these responsibilities added to the designer's obligation, the designer's role as leader of the design process becomes even more critical, since the interests of the environment and well-being of society become part of the paramount responsibility to safeguard the health, safety, and welfare of the public and environment. Perhaps, this is easier said than done.

The designer leads this process through the willingness of the stakeholders to participate. As said before, this is not an assembly of equals. The interests of the various stakeholders can be quite different, ranging from financial/economic concerns to quality-of-life issues. Added to these interests are the responsibilities that the designer brings to the table, argued here to include the interests of sustainability. The role of leader must be to seek consensus among the concerned parties and to synthesize a solution from the process. The leader must also be the voice for the future and environmental quality. In short, among the roles the professional must play is teacher. This obligation to pursue a design that contributes to well-being and flourishing requires the professional to inform the stakeholders of these aspects of the project and to communicate how their interests affect and are affected by these deeper aspects of the project.

The designer is obligated to make this effort and to pursue the outcome best for all interests, but he is not responsible for the actions of others. One can teach another, but you cannot make another learn. The duty to teach and to relate the need for sustainability is a necessary part of the professional's role, but is there a limit to the designer's obligation? If we step back to Alpern's Principle of Care and Corollary of Proportionate Care, the burden on the design professional is to prevent contributing to the harm that might be anticipated by a reasonable person acting with the knowledge and responsibility of the designer.[1] Our understanding of the implications of design as usual is that it is unsustainable and, in some cases, actually contributes to conditions that are defined as harmful. This creates a fairly high bar to measure the performance of the individual, in particular, and design itself, in general. The designer is an actor whose decisions, consciously and intelligently made, directly result in products and environments that directly and indirectly effect humans' well-being and the environment. This role places the individual designer in a role anticipated by Alpern's Corollary of Proportionate Care: a higher standard of care is warranted.

As noted, the obligation cannot be limitless; however, and, as suggested by McFarland, the Kew Gardens principles might serve as a guide for the designer to assess his own obligations as well as those exercised by others.[2] The preconditions of the Kew Gardens principles are meant to provide a framework for assessing a moral, but on our case, an ethical, obligation. The designer must determine in his own mind that these preconditions exist and what actions would a reasonable, knowledgeable person take. The first precondition requires that a critical need exists, that is, that there is something notable at risk or being threatened. In practice, how will we define the risk or threat to humans' well-being and the environment of design practices that we know are unsustainable? The next precondition requires the designer to be aware of the threat or risk, and the third precondition requires the designer to have the ability to help without harm to herself or others. The designer is in the unique position to understand and appreciate the issues of unsustainability and has awareness of the body of knowledge that exists to support the need for sustainable design. The designer finds herself in the position of being able to help. Finally, the last precondition of an obligation is that there is no other source of help. As leader of the design process, who else is there to advance abundance, health, flourishing, and growth? In the final analysis, sustainability becomes an ethical choice for the individual designer as well as for the design firm and the clients that enjoy their services.

It is the ethical obligation of the designer to lead the design process to a sustainable outcome, an outcome that contributes to the society flourishing. The obligation rises out of the special knowledge of the profession, the recognition of this knowledge, and the resulting sanction of the profession by the public and the professionals' consilient knowledge of other areas of knowledge. No educated person in contemporary society can claim ignorance of the impacts of human population on nature or the implications of unmodified growth and consumption. The duty to protect health, safety, and welfare of the public and the environment is paramount in the codes of ethics and canons of every design profession. Awareness of the obligation requires the design professional to embrace sustainability as an ethical duty. The difficulty rises in determining just what the duty requires of the designer. Since sustainability is difficult to define or to craft a definition that has real meaning for a designer, there can be legitimate confusion over what sustainable design is. As it is viewed commonly today, the duty is expressed in mitigating the impacts of a design when it is built, in essence, trying to reduce unsustainability. If the goal of sustainability is merely less waste, less pollution, and less unsustainability, how will we know what is enough? How can we assess our work? Is merely doing less damage and causing less harm sufficient to be labeled *"sustainable"* design? Sustainability must encompass more than merely reducing the impacts of design on the environment; it calls for a fundamental rethinking on the part of designers.

Standards Cannot Produce a Sustainable World

How will we be able to assess our work? There are many definitions of sustainability and sustainable design, but few are very satisfactory when it comes to actual applications. Perhaps sustainability is like quality: difficult to define but recognizable when you see it. Anyone ever involved with trying to create a meaningful definition of quality knows the difficulty in finding a description sufficient for broad application. Quality is complex; it is an amalgam of characteristics inherent in a product, process, or relationship and usually defies simple characterization. This seems to describe sustainability as well. What is sustainable will vary from place to place, from application to application, and from time to time. It will, however, be recognizable. For example, the Leadership in Energy and Environmental Design (LEED) process requires designers to meet certain tests in the design of a building to achieve certification. The expectation is that if these steps are taken, the building will function in a predictable, desirable fashion. The outcome is a result of the choices made by the designer and not a sustainable decision in and of itself. LEED certifies the expectation of an outcome.

In this sort of construct, sustainability becomes the result of a value system rather than emerging from prescriptive standards of engineering or material use. William McDonough and Michael Braungart have written about the need to embrace abundance and growth as the keys to sustainability. John Ehrenfeld writes that what we desire is a system that allows us to flourish.[3] Abundance, growth, health, and flourishing are the values of sustainability. The obligation of the professional designer is to lead the design process to outcomes that result in or contribute to abundance, growth, health, and flourishing.

One way of viewing the design process is as the process of avoiding failure; design proceeds weighing solutions in terms of what will work. Henry Petroski observes that design is always evaluated in terms of failure, and success is celebrated in terms of failure avoided. Failure is most often perceived as the antithesis of success, but it may be seen in several different ways. Failure can be a matter of perception; one person's failure is another's success—as in the old saying that the operation was a success but the patient died. We are all familiar with projects that were beautifully designed and constructed but simply failed to thrive. The project may still be recognized as a design success even if it has been an economic disappointment.

In other cases, failure is an event. Catastrophic failures occur in otherwise-successful designs. Unwise alterations during construction, poor maintenance, and unexpected uses might all result in the failure of a product or design that was otherwise a sound design. Success in cases such as this is a status that is irrevocably lost once failure occurs; once a thing is seen to have failed, it cannot be seen again to be otherwise. Success and failure in a design process or outcome may be defined in many ways. A design might fail economically because of poor financial planning, a change in the marketplace, unexpected costs, or simply the whim of the public. While engineering or structural failures are uncommon, they do occur; however, they are almost always limited to projects or designs that are on the cutting edge of design and materials. Catastrophic failure in other areas is rare primarily because of the standards of care and practices that have emerged, been tested, and been refined over long histories and are applied almost universally. Long experience has led to safety factors or practices of overdesign that are routinely employed to avoid risk and reduce liability. These safety factors and design practices have evolved from trial and error as much as from rigorous engineering and scientific study and are now so familiar to us that they are followed without question. Some design practices are so familiar that students learn the standard practice often without learning the principles or science behind it.

All responsible designs and designers proceed on the basis of what they have learned, either firsthand or as a student. There are very few truly inspired and new solutions to problems; solutions tend to be iterations of what has worked in the past. Failure may occur because the principle was improperly applied or because design problem has changed, requiring new use of a familiar method or an entirely new consideration. Still, as students or practicing professionals, we rarely study failure in any meaningful way. No one likes to dwell on mistakes, particularly one's own, but it is the lessons of mistakes that make our experience valuable. Failure is rarely discussed openly, and its causes are often dismissed as ineptitude, poor judgment, or misadventure. There are few resources, with some exceptions, of design literature that explore the projects that have not worked except as a gloss before moving on to other topics. The literature of site planning and design is primarily a catalog of practical methods and a library of success stories. Expositions on failure are uncommon, but we readily acknowledge that we learn more from understanding failure than we do from mimicking past successes. To study failure is not only to gain understanding of the underlying principles and the forces at work in a situation but also to gain an appreciation for the choices that were made and why.

Design failure is usually divided into technical and nontechnical causes. Technical causes are addressed directly in the practice of overdesign and safety factors. Most technical problems can be discovered in the quality assurance processes, primarily through repeated checking and multiple reviewers. As it happens, nontechnical causes of failure may be more difficult to identify or address.

If we look upon the products of design over the last 100 years, how will we access their impacts and effects? In the light of current understanding of the environmental and human health systemic effects of industrial manufacturing and the development habits of modern society, are these processes successful? Or, is there an argument for changes made to evaluate past practices?

In the past, we spoke of costs and benefits and strove to balance them, to maximize benefits, and to minimize costs. The analysis, however, was typically limited in scope to immediate economic concerns and firsthand implications. Indirect implications such as environmental costs and benefits have commonly not been included in the calculation. Much of our current environmental difficulty is directly attributable to these externalities; costs and benefits not considered in the original project. Designers looking for values of sustainability can visit designs of the past and study the true costs and benefits. Economists define economic efficiency in several different ways:

1. The maximum output achievable by a process without additional inputs
2. The maximum output possible at the lowest possible per unit cost
3. Maximizing the satisfaction of at least one person without decreasing the satisfaction of another person

Much of our economic success has been measured using the first and second of these definitions. The third definition is used and discussed less often. The values of sustainability require that this third definition be employed first and that it be broadly interpreted, in a sense, as a goal of design efficiency. Designers must think in terms of outcomes for the users, for the environment, for the future, and for the clients in order to achieve a goal that maximizes the satisfaction of the various stakeholders without decreasing the satisfaction of any stakeholder.

The design process then is necessarily performance oriented, as opposed to merely following guidelines and rules. Meaningful design prescriptions are as unlikely to produce desirable outcomes in the future fundamentally better than they have in the past, because they are based on unsustainable assumptions and principles. Instead, the professional designer must employ his or her training and special knowledge to achieve ends that are consistent with or reflect these values. Synthesizing a design that will contribute to achieving the values of sustainability will require more from the designer, from the reviewer, from the client, and from the authority.

Designing by standards is designing to the minimum; designing to achieve a sustainable outcome requires more. This requires careful attention to detail; understanding the nuances of the project at hand; considering the systemic impacts of design, the costs and benefits of material choices, and issues of welfare and flourishing; and avoiding results that limit these values. Since few activities, perhaps especially the activities of the modern human, are without impacts, it becomes critical to begin to manage the impacts and create designs that account for them, to appreciate the closed loops of material flows and open systems of energy income, and to incorporate that appreciation into design at every level. The broader knowledge required for successful designs will likely result in coalitions or collaborations of knowledgeable people. Standards of performance will emerge as our experience grows, but prescriptive standards so prevalent now early in the development of sustainable design will not produce the desired outcomes. Clearly, some standards that refer specifically to health and safety and that have emerged from the process of trial and error and careful analysis will remain, but more prescriptive standards that deal in form rather function may not.

Instead, the standards of a sustainable design process will be performance standards that do not dictate what a designer shall do but what outcomes are desired. The design process becomes engaged in how to meet those performance objectives. Design success and failure will be assessed in terms of how well the design performs in context of the desired outcomes and probably indirect benefits as well. The performance standard approach maximizes the flexibility and creativity of the designer. The potential for innovation is increased, but so are the challenges. There is greater risk for everyone in an environment where performance standards are used. The designer is at greater risk because she may be working without the comfort zone of prescribed design elements.

The regulatory bodies and consumers may have greater risk because, by using performance standards, they are regulating how the design functions and interacts with existing products but not how the new element is fashioned or formed. For example, issues of community density might be managed in a variety of different ways to meet performance standards of open space or traffic management. A community designer might choose high-density clusters or uniform density with smaller green spaces, a combination of these, or something more novel. So long as the performance objective was met (open space), the community would have little choice but to accept the design. Likewise, the clients may have to accept more risk for the flexibility provided by performance standards. In a performance-based system, most of the design work may be completed quite early in the review process, in some cases even before the review process begins. This means little or no feedback from regulators and users before a good deal of design expense is spent. There is more risk in a performance-based approach for all parties, and so, there is likely to be a tendency to create and rely on standards at the expense of true sustainable outcomes.

Many design professionals are employed in the review of the work of other professionals. They have a unique role in the design process in that they are expected to share the special knowledge of the professional, but they represent another stakeholder besides the project or the client. This stakeholder is likely to be a regulatory agency or government with some regulatory or permitting authority. The reviewer may be an employee or a contractor. Since the reviewer is actually not engaged in design, what ethical obligations do they have for the final product? The reviewer may have significant influence on the outcome of the design; changes are made, often significant changes, on the basis of the reviewer's work. The regulatory agencies relies on the expertise and professionalism of the reviewer to represent their interests and to assure compliance with standards. Quite often, the reviewer works for supervisors or officials who do not have the expertise to evaluate the technical aspects of design, even though they may be empowered to pass judgment

The Choice for Sustainability

on it. The reviewer then serves several stakeholders; clearly, the employer or client and the profession and its ethical obligations is among them, but paramount is the duty to protect the health and safety of the public and to promote their welfare. To the extent that changes in a design are required by the authority based on recommendations from the reviewer, the reviewer is engaged in design and shares the ethical obligations for the outcome of the design. Were it to be otherwise, the reviewer would in fact be directing the designer without any ethical or professional responsibility for his work and its implications.

Reviewers are sometimes not qualified to actually do the work that they assess. They do not necessarily hold professional licenses, for example, but instead review on a comparison basis. In these cases, the proposed project is compared with a list of standards acceptable to the reviewing authority. Items that are in variance with the standard are required to be brought into compliance or receive a waiver to be accepted. By definition, to achieve as broad an application as possible, a standard of any sort must seek the lowest common denominator to describe what is acceptable. The standard is, in essence, a design without analysis; there has been no specific assessment of the problem to select the appropriate solution. Instead, the standard calls for the designer to select from a predetermined solution or set of solutions to be used in all cases of a certain type or condition. It is the designers' role to make them work rather than to determine the best solution. It is probably true that standards save clients' money on occasion and provide the designer with at least a nominal justification or defense for the design elements that are selected. One must only look around at the built environment, the problems caused by uncritical application of standards to realize that the use of design by prescription has not worked. It is no leap at all to realize that sustainability "standards" are unlikely to be useful or achieve the values that define sustainability. Designers working toward sustainability necessarily are committed to designs that achieve the greatest possible contribution to abundance, growth, health, and flourishing. Sustainability is deeper than reducing unsustainability and resists "cookie cutter" approaches. The only meaningful standards then become performance standards, and the question becomes how we will gauge whether a design contributes to a sustainable prosperity.

Design is about making our intentions manifest in the world. What are the intentions displayed by products and environment that we have designed? The use of toxic materials and processes that consume energy and materials far beyond the value of the products produced, the creation of wastes that continue to have negative effects long after the products are consumed and disposed of, the production of food that is inherently unhealthy to eat and of neighborhoods that contribute to poor health, and so on, all suggest intentions that are unconcerned with health, growth, abundance and contributing to our combined prosperity. What are the intentions of this design? If design is merely the thoughtless reiteration of what we have already done and of what already exists, then we have a clear understanding of what the future holds.

The idea of sustainable design is to break with the ideas that have produced a body of work that is now understood to not work in the long run, to not contribute to health and prosperity of society, to consume nature for short-term advantage, and to have little regard for the future. In this system, only economic capital has had value, and nature appears on balance sheets only as an expendable resource. Our awareness of understanding in other fields is sufficient to say that this short-term thinking is inadequate. For the designer, it means defining concern with health, safety, and welfare as limited in scope to short-term considerations of only the most direct and immediate effects of his work. Longer-term implications are the responsibility of "others," such as reviewers, clients, regulatory authorities, stakeholders, and government. What of the special knowledge that distinguishes the professional designer? What of consilience? How does the professional proceed?

Sustainability and Obligation

In the final analysis, the designer is the leader of the process, but he is not the only participant. The designer is obliged to think through immediate risks to the public's health, safety, and welfare and to address harm that might be reasonably anticipated. The farther out the horizon for those considerations, the greater the uncertainty that accompanies them. The designer may be aware of the potential for harm in the future or from some indirect impact consequent to the design, but these concerns may not rise to the level of being issues for other stakeholders. In such an event, the designer would have to make a judgment regarding the ethical obligation to prevent or avoid some distant future harm in favor for a design benefit closer at hand.

The problem may arise in the individual's mind if she believes that making the ethical choice might be expected to lead to the loss of one's job or professional opportunity. At what point is an ethical act no longer an obligation? The bar for the professional is high; the paramount duty is to protect and promote the safety, wealth, and welfare of the public and the environment. Should an individual designer be faced with such a difficult choice, he must act in the public's interest, unless such a choice has consequences that outweigh the public's interest. These consequences, however, require balancing the interest of the individual designer against the combined interests of the public.

A more likely scenario is the designer working within a process in which she is the leader but not necessarily the sole decision maker. It is the duty of the professional to represent the interests of public health, safety, and welfare and, as previously argued, the interests of sustainability to the stakeholders in the process, but it may not be within the power of the professional to convince all players. If the designer's recommendations are not accepted by the stakeholders or are overruled by a regulatory authority, has the obligation been met? In such a case, the nominal ethical act is for the professional to protest the decision and to present the facts and his recommendation once again before accepting the ruling. The ethical requirement may not exist in the presence of such a law or ruling if the professional is no longer able to act (third precondition of the Kew Gardens guidelines). It may be unpalatable to acquiesce to such a ruling, but the professional also has a duty to obey the law. Even if the professional should be aware of the imminent harm, the ethical duty to prevent it would exist, but the power to act may be precluded.

In the absence of such a ruling or law, the duty to prevent harm exists. A distinction should be drawn here between acting to prevent anticipated harm and a duty to do no harm, as previously discussed. A promise to do no harm creates a threshold that may not be possible to maintain. In the interaction between stakeholders and forging a consensus in design, it is possible, even probable, that some parties will feel "harmed." It is unavoidable as part of the stakeholder process and human nature. The actions and decisions of the designer must be guided by something else, maximizing the utility of the decisions for all affected, including the interests of nature and the future. This brings us full circle to back Chapter 3 and the short discussion on whether actions should adhere to act utilitarianism or rule utilitarianism. Act utilitarianism considers an act right if and only if the good or benefit of the act outweighs the bad of the act. Also, good and bad are not determined solely in terms of the actor but for the benefit/harm of all. It requires the moral agent, in this case the design professional, to think of the consequences for all. Rule utilitarianism, on the other hand, distinguishes between an act and a rule. Rule utilitarianism requires that moral rules be observed and that these rules be followed based on the greatest utility or tendency to promote happiness. In the end, the individual must decide what is the proper course of action and be aware that the decision will be held to the professional standard of care and what a reasonable, knowledgeable professional would do. Where is the greatest happiness for all most likely to be found? Clearly, the public and stakeholders rely on the professional's special knowledge and ethical judgment to produce the best outcome, whether guided by an act ethic or rule ethic.

If the professional finds that the proposed actions do constitute a threat to health, safety, and well-being of people and the environment, the duty to prevent this clearly exists. In the absence of a situation where the designer is powerless to prevent an act, there is a paramount obligation to protect and promote the health, safety, and well-being of people and the environment. Where the individual designer is constrained from doing so by contractual relationships, supervisory interference, or other sublegal influence, the ethical obligation to protect health, safety, and welfare is paramount and exists. There is no relief from this duty. Further, the more critical the risk or threat, the greater the obligation to act. In such a case, the professional could choose to withdraw from the project rather than proceed. The risk of harm is uncertain or minimal when compared with some acknowledged and expected benefit, and the professional might choose to proceed, having made the concerns a matter of record.

Once the duties of the professional are understood in the broad strokes, though, it becomes clear that the obligation to act sustainability cannot be limitless. By virtue of the many interests involved in a project, the benefits and costs of design are unlikely to be distributed uniformly. Sustainability in the end is a balancing of interests. Every act has consequences; the implementation of every design involves materials and energy, consumption of resources, impacts on the future, and exchanges of value here and now. Furthermore, our understanding of costs and benefits and of implications is incomplete and constantly under revision. One obvious limits of the obligation to act sustainably or to protect the public health and safety is the professional's right to rely on what is known to be true at the time she acts. In any given career, there are instances where decisions made in the early years of practice bear little in common to decisions of the same nature in later years; as we gain experience and our understanding of our work and world changes, our decisions change. As stated earlier, the standard of care is and must be dynamic. Likewise, failure to comply with evolving standards of care may make past successful approaches to a design problem unacceptable in future instances.

As already noted, designers do not work as solitary actors but within processes composed of stakeholders, reviewers, standards, statutes, and protocols. Further, the professional is responsible for maintaining a working knowledge of his area of expertise, not merely in the context of materials, methods, and design but also in terms of the implications for public health and safety and the environment. This obligation requires some effort at consilient knowledge; what is being discovered and learned in other professions? For example, there is some academic research that indicates that there is a correlation between student learning and sunlight. Should we expect designers of schools and classrooms to be familiar with this research? Would a classroom designed to employ such information reflect a deeper appreciation for the welfare of the student stakeholder than the one that did not? Of course, quite often, research findings are undone in subsequent studies; perhaps waiting for corroborating data is a better service to client stakeholders. When does new knowledge acquire sufficient weight to be an issue of concern for the professional?

Science and Design

Science plays an important role in design. In one sense, science provides the foundation for many design decisions. The characteristics of materials, how things behave in nature, and how we interact with the products of our design are among the areas in which designers rely on and employ scientific knowledge. In some ways, science precedes design as well. The growing awareness of global environmental crises and the implications of a burgeoning world population are largely the result of a deeper scientific understanding of the relationships between human activity and natural processes but also the realization of the surprising fragility of nature. Science has provided us with a deeper and richer understanding of nature and our place in it than was anticipated even a few decades ago. This awareness and understanding has led to increasing calls for a new design and cultural paradigm, but with these calls come challenging questions.

Science, like ethics, is both a body of knowledge and a set of rational processes. We often rely on science knowledge, without appreciating its inherent dynamism. On the other hand, calls for decision making based on scientific certainty do not appreciate either the inductive nature of much science or the likelihood that tomorrow will reveal nuances unsuspected yesterday. The role of science in the design process is to provide designers with the ability to determine the degree of uncertainty in our knowledge as much as to reliably predict the behavior of materials used in design. Engineers are familiar with margins of safety built into products and structures. One way to view a safety factor which calls to overdesign a thing two, three, even four times is as a measure of uncertainty. It is important to recognize the limitations of science and the absence of absolute "proof" when it comes to design applications. Still as much as science might require us to be skeptical consumers of information, it remains the best source of usable real-world explanations for phenomena in the real world. Science is a skeptical and reiterative process, but designers must manage uncertainty in their work and rarely get an opportunity to do precisely the same work again. Designers require information that is of sufficient quality to be acted on, to be used to inform decisions. When is there sufficient weight in science to consider findings robust enough to act on?

Science can be viewed as method, as in the scientific method, or as process that includes considerations beyond the scientific method itself. The work of Kuhn and Popper, among others, clearly demonstrates that science as a process differs substantially from the science anticipated by the method alone. Experimentation remains an important part of science, but in the actual practice of science, it represents a relatively small share of the activity. Further, what experimentation that does occur is about falsifying existing knowledge as opposed to testing or creating really "new" knowledge. In the end, most scientific activity is involved in developing explanations for what has already been learned by observation and experiment.

Once scientists have collected their observations, they have to determine what those observations mean. This requires the scientists to draw inferences from the data based on logical reasoning. Deductive reasoning is logically valid. It is the fundamental method used in mathematical proofs. Deductive reasoning proceeds from the general rule to the particular, whereas inductive reasoning proceeds from the particular to a general rule. In inductive reasoning, we say a general principle is true because all of the examples or samples we have seen are true. Of course, inductive logic is not logically valid, because the limited number of observation is insufficient to guarantee a general rule. But inductive logic is consistently used in science and even in day-to-day life.

Inductive reasoning involves drawing a general conclusion from a limited number of observations or results. The classic example is of observing crows. If the observer notes that all of the crows she has seen are black, she might reason that "all crows are black." Indeed, she can use this hypothesis to predict that the next crow she will see will indeed be black and, more than likely, it will be. In fact, she might go on for the rest of her life observing crows and never see a nonblack crow. Each observation strengthens her hypothesis that all crows are black. The problem of induction is that, no matter how many observations she makes, she will never observe all crows, and so, her hypothesis cannot be proved. The incidence of one nonblack crow would prove her hypothesis wrong, falsifying her hypothesis. Induction is not a logically valid method of reasoning. There is no mathematical proof for an inductive claim.

Since most scientific knowledge is achieved through induction, uncertainty is a necessary reality. Since no practical amount of experimentation or observation can overcome the logical problems of induction, credible scientists are generally careful to craft their language in such a way that will account for the limitations of the inductive process. This professional language can be baffling to nonscience consumers of data. This appearance of uncertainty suggests that some sense of skepticism for science is appropriate. Much of the work of scientists is looking at the work of others and evaluating the conclusions drawn. This activity involves looking for alternative explanations and following the logic of the original proposition. How does the information comport to existing understanding? Is the explanation or observation consistent with other similar work? Is there a logical thread that leads through the experiment to the explanation? All outcomes and explanations are subject to critical evaluation and falsification.

All criticism, however, is not equal. Ideas and explanations are not quickly abandoned in the face of just any alternative explanation. Alternatives themselves must meet the same tests of the original hypothesis; that is, they are subject to critical assessment and falsification. The hypothesis may survive the challenge, by adapting to either include or falsify the alternative explanation. Sometimes, mutually exclusive explanations stand until one is falsified or a new explanation is able to resolve the differences. Science is not always as tidy as one might want to believe. Continued attempts to falsify a hypothesis add to its credibility. While no tests can prove the idea to be true, over time, a hypothesis may have survived sufficient attempts to falsify it that it becomes largely accepted as a good workable explanation, and it is accepted as a theory. In essence, it is treated and used as if it were "true."

If a hypothesis or a theory is an accurate explanation of some phenomena in nature, it can be used to make predictions about the world. Scientists can use the theory to predict or anticipate some natural phenomena and then set about to see if the prediction is true. Again, no amount of successful prediction is in itself "proof" that the hypothesis is true, but as success accumulates, the theory is viewed with more and more credibility.

Since objective proof in not possible in most cases, science proceeds on the basis of demonstrating things to be false. Science as a way of knowing is based on facts established by observation. It is often said that science is the search for the truth, but because of the limitations of observation and understanding, what we really accept is the best explanation for observed or inferred phenomena in nature, as opposed to a fixed immutable "truth." The scientific method is a process, a method of inquiry that involves posing a question, making observations about the question, analyzing what is seen, and then drawing a conclusion. The conclusion is frequently incomplete, and the analysis gives rise to new questions rather than complete answers. Science is *inconclusive* to the extent that there is always more to be known.

A key strength of the scientific method is that it is repeatable. We can build on the work of others without necessarily repeating it. In science, it might be said that we know something until someone undoes it, and then, we know it in a new way. Laws of science are those things that are, or we believe to be, immutable, that is, unchangeable, and always true. These are things that are known deductively. Everything else is theory or hypothesis. In the 1930s, Karl Popper attacked the use of inductive logic in science. Popper's work required scientists to adopt the principle of falsifiability in order to justify general principles based on inductive logic. Though paradoxical, falsifiability is an important concept in science that requires that a theory cannot be scientific if it does not permit the possibility of its being false. This is critical. Falsifiability does not mean the theory is false, only that scientists allow, in principle, for the possibility that an observation could occur that would show it to be false. Any theory that is not falsifiable is not scientific.

To be scientific, observations must be repeatable; that is, the experiment or field observation must be able to be recreated or witnessed by others. In 1989, two chemists at the University of Utah claimed to have created cold fusion in a desktop experiment, but they, and many others, were never able to repeat the experiment successfully. The fact that their observations were not repeatable caused the idea of desktop cold fusion to sink fairly quickly from the scientific horizon. Another important aspect of science is that observations and conclusion must be transmissible; that is, they must be able to be communicated from one person to the next.

Hypotheses are questions posed about natural phenomena. Hypotheses must be testable; that is, the question must be framed in such a way that it can be tested. The questions must try to minimize the number of possible explanations and influences. Theories in science are explanatory propositions, that is, ideas offered to the scientific community as explanations for things observed in nature. Theories emerge from the process of testing hypotheses. Theories become accepted over time as scientists test alternative theories to see which might be a better explanation. Over time, a consensus emerges that the theory is or is not the best possible explanation of the facts. When science has tested a hypothesis sufficiently that the theory is generally accepted as true, sufficient to describe it as a theory, one might say at that point that the data is robust enough to begin to use it to act on.

It is clear that different people will determine when the evidence supporting a hypothetical gains sufficient weight to be considered a theory and therefore worthy of our consideration when planning and designing. One must look only to the various opinions concerning global climate change over the last 20 years to follow that process. It is necessary to acknowledge that these processes of acquiring new knowledge and beginning to act on it are themselves uncertain and gradual. Each step of the way has its own lag time. The process of science is based on skeptical consideration, reiterative evaluation, and, if appropriate, gradual acceptance. Once there is a general acceptance of the science, the movement from study to application is also slow. The process of scientific revolutions described by Thomas Kuhn probably has a corollary in design, wherein innovations emerge from younger designers, with less invested in prior work, knowledge, and success, than their older colleagues. If this is true then we might expect older, perhaps I should say, more established designers to accept a new design paradigm more slowly than younger designers still establishing themselves in their field.

Sustainability and Design Ethics

Design differs from science in several important ways. Key among the differences is that science is unable to provide answers to questions of value. This is not to say scientists do not have values because, clearly, that is not the case. Rather, science in its attempt to be objective is typically unable to make value judgments about knowledge without slipping form the objective viewpoint. So, while scientists may have values, they are generally not part of the scientific paradigm. There are value-based aspects of scientific practice, ethical standards for the treatment of subject in studies, for example, but the data is not, at least in theory, filtered through values.

Design is entirely different from science in this way. Design proceeds based on the values of the designer and the stakeholders. When we evaluate design, we bring to the process our own values. There is a design philosophy at the heart of every design effort. Just as scientists need a grounding in their area of interest, to know the history of the field and to understand the methods used and how discoveries have shaped the current science, designers are inculcated in the values of their profession. Design schools prepare future designers by instructing them in the history of their field and by teaching about materials and how they are used, methods, esthetics, safety, and so on. Students bring their own values and experience to their lessons and build on these and their education, as they proceed through career. At the heart of the design process are values of beauty, of environmental and personal integrity, and of service and intellectual accomplishment. These constitute the design philosophy of the individual and the ethical construct of the professions.

Designers has obligations that no other stakeholders in the design process have. They have expressed ethical duties to protect the health and safety of the public, to provide for and enhance the welfare of the users of their work, and to safeguard the environment. Regarding the latter, either by an ethical duty within their professional canons or by inference, can stakeholders remain healthy and safe in circumstances of continued environmental decline? No other stakeholder or participant in the design or approval process carries the weight of these obligations directly or to the extent of the design professional. Clients and users rely on the professional to avoid future claims of harm or negligence. Reviewers, even those from the design professions, serve to assure compliance with standards, rules, or regulations and the interests served by the reviewing institution. Other stakeholders are concerned with their own interests that may include health, safety, welfare, and environmental concerns, but frequently if they do so, it is at the exclusion of all other concerns, often as a surrogate for other issues. The interests of the future are difficult to know with any certainty and to express in meaningful ways. The environment itself is mute. It is the designer's responsibility to synthesize an outcome that balances the needs of the stakeholders, the environment, and the future. In light of these influences and constraints, how is it that the professional can balance the interests of all of these parties as well as the environment and the interests of generations yet to come? What is it about the education and the experience of the professional that enables such important and profound consideration?

The designer has the obligation to make a determination regarding public health, safety, welfare, and the environment. This obligation requires the designer to seek an understanding of what the implications of her design are and to propose designs that promote public health and welfare that enhance environmental quality and that are concerned with the ultimate disposition of the project. This requires a knowledge of issues beyond design alone. An understanding of what is being learned in the various sciences, natural and social, is necessary. The designer is only one participant in the design process, but she is the one with the greatest responsibility and deepest obligations. These are inherent in the nature of being a professional. Design, though, is a process and one with shared responsibilities. The designer's obligation is to promote outcomes that are consistent with flourishing, as described earlier. If these proposals are met with reluctance, it remains the duty of the professional to promote them and to act as an advocate and an educator. If science exists to the degree that a health, safety, and welfare argument can be reasonably made, the obligation exists for the professional to incorporate them into his thinking.

Clearly, in areas of uncertainty, the professional obligation would have limits. If there is sufficient uncertainty to question the designer's recommendations, a judgment must be made on whether to revise the proposal or design or to defend the proposal. Among some of the professional canons and ethics exists an option that the professional's obligation may be fulfilled by simply notifying the disagreeing parties that this recommendation is made within the purview of the professional's judgment and revisions are made with this protesting caveat. In such a case, the professional would be deemed to have met his obligation. It would remain to be seen if the professional community would agree and consider the standard of care met. Even where the obligation is clear, some uncertainty about our design choices will remain. We must acknowledge that our ability to predict the future is woeful. Even short-term horizons are difficult to see clearly. Faced with an ethical obligation to consider the interests of the environment and the future and to produce outcomes that contribute to human well-being, it is necessary for the ethical design professional to proceed and lead the design process carefully. The underlying science that supports design is by definition subject to change as new information as more is learned and as testing proceeds. Still, while doing so, the design must rely on information and, at the same time, acknowledge a degree of uncertainty. In the face of expected uncertainty, how is the responsible, ethical designer to proceed? It requires precaution.

The nature of the scanning required of the designer to be aware of the probability or potential for harm cannot be limitless, and it cannot be beyond reason. The designer must be able to rely on common accepted knowledge when making these assessments. For example, knowledge pertaining to global sea level rise has changed considerably over the last several decades. Even though the levels of sea level rise expected in the future were within the ranges published in older studies and reports, they were generally considered to be of less probability than they are now. Should the engineer designing port facilities in 1980 have planned for a sea levels to be more than 2 feet higher over the life of the facility on the drawing board? It is unlikely that his peers would have considered that to be necessary under a reasonable standard of care. The science was there in an early, less corroborated form, insufficient to act on. Would that same engineer designing a similar facility now be able to make that same decision? Unlikely. Still, many coastal communities and states have not revised their building codes or planning in ways that reflect the current anticipated sea level rise. Can a prudent designer then rely on such regulatory structures to guide design decisions made today? The question is made more difficult for some professionals whose peak organizations extend the duty to prevent harm to property.

In the end, it is always the same: the professional has obligations that require more than mere compliance but are not without limits. These obligations demand the individual designer to be aware of the development of new knowledge as it might influence her design and to anticipate harm in a broad sense using existing science and other knowledge. The nature of knowledge and risk are that they are not fixed and immutable but are uncertain. The purpose of the standard of care as a measure of the appropriateness and quality of decisions reflects this inherent uncertainty and demands that design decisions be consistent with what a reasonable knowledgeable person would do in similar circumstances. This is an uncertain measure in itself but one that accommodates changing knowledge and reflects that in the expectations of professionals. This circumstance however may face very real and fundamental challenges as design by means of artificial intelligence (AI) becomes common. Requirements of the designer of AI will require no less concern for sustainable outcomes than the more traditional designer. Now we must train an independent agent, a design in and of itself to weigh the considerations of equity, fairness, environmental sustainability as well as the parameters of the thing being designed. At least at this point no one is making arguments that the AI designer is a moral being and so the designer of the program has ever greater responsibility for the learning and thinking product they bring to the world.

Endnotes

[1] Alpern, Kenneth D., "Moral responsibility engineers," in Johnson, Deborah, editor, *Ethical Issues in Engineering,* Prentice-Hall, Englewood Cliffs, NJ.

[2] McFarland, Michael C., "The public health, safety and welfare: An analysis of the social responsibilities of engineers," in Johnson, Deborah, editor, *Ethical Issues in Engineering,* Prentice-Hall, Englewood Cliffs, NJ.

[3] Ehrenfeld, John R., *Sustainability by Design,* Yale University Press, New Haven, CT, 2008.

6

The Precautionary Principle and Design

A Thing is right when it tends to preserve the integrity, stability and beauty of the biotic community. It is wrong when it tends otherwise.

Aldo Leopold

Faced with the competing stakeholder interests, the uncertainty of predicting outcomes of design, and the demands of working in a competitive marketplace, the designer must find ways to balance these concerns with the ethical obligations of his profession. The concept of the standard of care provides the working professional with some measure of how one's decisions and work might be evaluated and assures him that, should it be necessary, excepting negligence, an evaluation of his work would be compared with what his reasonable knowledgeable peers might do in similar circumstances. The standard of care then becomes the process by which prudence and risk in design are assessed.

Ultimately, the standard of care is predicated on the fact that a duty of care exists, as has been described throughout this book. This duty places certain obligations on the professional, which in turn are addressed by how the special knowledge of the professional is applied. Since bodies of knowledge are not static, the professional must be engaged in career-long education in the facts salient to her work and maintain an awareness of how these facts might influence the scope of her professional obligations.

We live in challenging times. The total amount of knowledge is said to have doubled in the last 2 years, and the rate of new knowledge suggests that the current doubling time is 18 months.[1] Of course, this means that half of what is known now was not known a year and a half ago. Even if these numbers are not completely accurate, it is clear that the amount of knowledge and data available today is significantly greater than it was not so long ago. Most of this knowledge may not be relevant to the typical designer, but much of it may be. Industrial designers of plastic items should keep abreast of the growing scientific concern of bisphenol-A (BPA), for example; should the scientific data reach a critical mass, BPA may be banned from products in the United States, as has been in other countries.

The designer must rely on reputable and nonbiased sources of information to be able to assess the quality of information and the degree to which new knowledge has compelling science or reasoning behind it. In the end, the designer is required to act prudently to assess and mitigate risk in his work. The precautionary principle (PP) describes an approach to this that might guide the designer in accommodating new knowledge to meet her obligations and that would be likely to meet the test of what a reasonable, knowledgeable person might due.

It is difficult to identify with any precision where the PP first took form as an articulated concept, but the idea of caution is as old as nature itself. Precaution implies forethought or anticipation, a choice where alternatives are distinguished from one another by a degree of risk or perceived hazard. It is a choice between two different acts rather than a choice between inaction and action. As a defined concept in public policy, the PP is most often described as emerging in Germany in the 1970s. This is believed to be a reflection of a growing concern with environmental quality. In the years that followed, it was incorporated in a number of international treaties and agreements, most notably the Charter of the European Union and the United Nations Conference on Sustainable Development.[2]

There are any number of definitions for the PP; however, the most common one emerged from the Wingspread Conference in 1998:

> "When an activity raises threats of harm to human health or the environment, precautionary measures should be taken even if some cause and effect relationships are not fully established scientifically. In this context the proponent of the activity, rather than the public, should bear the burden of proof."[3]

Another key definition is the language incorporated into the Maastricht Treaty that laid the foundation for the European Union. Article 130R(2) reads:

> "Community policy on the environment shall aim at a high level of protection taking into account the diversity of situations in the various regions of the Community. It shall be based on the precautionary principle and on principles that preventative action should be taken, that environmental damage should as a priority be rectified at source and that the polluter should pay."[4]

The PP emerged out of public policy in the 1970s and 1980s in Germany in an attempt to prevent environmental damage by stopping it at its source. The Maastricht Treaty, which created the European Union, included the concept among its precepts, saying in Article 174 that "Community Policy on the environment shall aim at a high level of protection. It shall be based on the precautionary principle and on the principles of preventative action should be taken, that environmental danger should as a priority be rectified at source and the polluter should pay…"[5] In essence, environmental damage should be prevented before it occurs. In its nascent form, it was intended to provide a basis for public policy to keep up with or in advance of scientific knowledge and its commercial application. It was first formerly applied in the First International Conference on the Protection of the North Sea in 1984.

Since then, it has been included in a number of international agreements and treaties. Although there is no universally agreed-upon definition, the PP is commonly explained as "When an activity raises threats of harm to human health or the environment, precautionary measures should be taken even if some cause and effect relationships are not fully established scientifically. In this context the proponent of an activity, rather than the public, should bear the burden of proof."[6] The basis for the PP is a simple one: our experience with modern technology has included numerous unintended consequences; in the future, we should make every reasonable effort to anticipate and avoid such consequences.

In the past, the approach to scientific, technical, and commercial activity or innovation has been to introduce a material, product, project, or process and, after the fact, to deal with or ignore the negative impacts. In essence, the acceptable practice has been to react to negative impacts only after the damage has been done. Of course, in many cases, the damage cannot be reversed and penalties such as fines are commonly inconsistent with the extent of the environmental, human health, and economic impacts. Such an approach subsidizes economic activity to the extent that the public and the environment bear the costs, while stockholders and owners reap the benefits of externalities by avoiding the costs. Further, it has most often been the burden of those affected rather than those responsible to prove the cause and effect of these impacts.

Moral Underpinnings of the Precautionary Principle

Among the literature, there is wide acknowledgment on both sides of the issue that the PP is a values-based concept. A definitive analysis of it has not been completed largely because of there is no clear nomenclature and/or agreement as to what it precisely is or how it is to be applied. It is not surprising then to find that there is actually very little written about the moral reasoning in support of precaution as a principle. Such a concept stands in stark contrast to the market-based approaches of the past, where a product or a service was introduced and the public bore the risk and cost of determining the risks or hazards associated with it. Product safety most often was a function of imposed regulations, economic risk assessment of the costs of liability, or market pressure of various forms. The PP stands the principle of *caveat emptor* in its head insofar as environmental impacts are considered. Predictably, such an effort is controversial.

Christianson has observed that in its current form, the PP embodies six principles[7]:

1. Preventative anticipation: "a willingness to take action in advance of scientific proof of evidence of the need for the proposed action on the grounds that further delay will prove ultimately most costly to society and nature, and, in the longer term, selfish and unfair to future generations."[8]

2. Expanding or improving the assimilative capacity of the environment.
3. Proportionality "of response or cost-effectiveness of margins of error to show that the selected degree of restraint is not unduly costly. This introduces a bias to conventional cost benefit analysis to include a weighting function of ignorance, and for the likely greater dangers for future generations if life support capacities are undermined when such risks could consciously be avoided."[9]
4. Obligations to care, that is, creating a formal duty of environmental care. This puts the onus of proof on proponents of change. It is this feature that is often criticized as limiting innovation and risk taking. On the other hand, some see the principle as a source of inspiration and innovation.
5. Promoting the cause of intrinsic natural rights. An expansion of the legal notion of ecological harm is widened to include the need to allow natural processes to function in such a manner as to maintain the essential support for all life on Earth. The use of ecological buffers in future management gives a practical emphasis to the thorny ethical concept of intrinsic natural rights.
6. Paying for past ecological debt.

Still, the PP is not without its critics and challenges. Very often, the evaluations necessary to identify an impact must be made without substantive information. The real impacts of a product or activity may simply not be known or anticipated until well after the fact. Critics of the approach assert that such a burden will chill, perhaps even eliminate, technical, scientific, and economic innovation. Experience would indicate that anticipating impacts is very difficult. Even statistical analysis is often insufficient as a predictor of outcomes in cases of new technology or science.[10] As it is commonly framed, the principle calls for a reasonable assessment and caution to be observed, even when the science is incomplete or the degree of uncertainty is high. Critics have observed that, from this perspective, we would be eating all of our food raw because of the unfavorable risks of using fire.[11] There is something to be said for this criticism: we often received unanticipated benefits as well as negative impacts from economic development and new products.

The practical use of the PP presents a number of challenges. Critics often predict that broad use of the PP will result in the end to innovation,[12] the end of science,[13] redistribution of global wealth,[14] and a race to the bottom.[15] Most of the printed word on the PP is committed to arguing for or against the use of precaution in a particular, usually narrowly defined, case. For example, much is written about the application of the PP and genetically modified (GM) food or precaution and global warming. These specific arguments generally share one or more basic criticisms. The key arguments leveled against precaution are that it is not scientific and not coherent and that it has its own risk of unintended consequences or that it will be the end of progress.

Morris provides a good analysis of the PP. He notes that, in general, the definitions fall into one of two fairly broad definitions that he call the strong PP and the weak PP. Strong PP requires that no action be taken unless there is certainty that it will do no harm. Weak PP states, on the other hand, that full certainty that no harm will occur is not required to proceed with an action.[16] Strong PP is an unworkable standard, since certainty is not possible and requires proponents to prove a negative.[17] Weak PP, on the other hand, requires precaution and perhaps a shift in the form of discussion from objective data to politics or values.

At the center of the discussions is the idea of our ability to anticipate and calculate risk or negative implications of an act. Risk assessment is usually discussed as a completely objective undertaking, but, in fact, every risk assessment contains within it assumptions and limitations that reflect our limited ability to model anything as complex as the real world and the future. These assumptions are very often value-based assumptions that reflect the beliefs and bias of the assessors. The degrees to which these nonobjective elements influence the outcomes of risk assessment vary, sometimes significantly. The algorithms of contemporary risk assessment may have the patina of objectivity but not the substance.[18]

Claims of or calls for scientific certainty necessary for strong precaution are always hollow. In the end, perhaps, decisions about the world and technology are not questions of science but questions of values. Because we are able to do a thing, does it follow that we should or must do it? If this is the case, the question is not whether decisions are to be based in science or politics but how we consider the former in the process of the latter. How then might science be modified to serve more precautionary purposes? Cranor begins by asking, "given what we know at present, what could precautionary science be?"[19] He notes precautions are routinely taken to protect things that we value. We plan and organize in anticipation of a threat or risk to protect these things: our children, valued possessions, etc. It is key that we do not wait for the actual threat or harm to occur and then respond but that we think and anticipate to avoid the threat. He suggests that if the environment or resources are considered valuable, it is likely or possible that we would eventually develop corresponding anticipatory methods. The emergence of the PP as an articulated concept in public policy may reflect such a shift in values.

Design is an amalgam of art and science. Science in principle is usually described as an objective search for the truth in nature. Modern science, however, is conducted in laboratories funded by corporate and other interests. Those interests determine the foci of research, and those interests interpret the results. In light of these trends, science cannot be viewed as not having a point of view or goals beyond merely the "truth." It has contributed much to our understanding of the environment and also much to "This paradigm of nature-as-capital, reducible to fungible, transnationally tradable units of biocurrency..."[20] Science has important contributions to make, but ultimately, it has no capacity for decision-making and no inherent values, and it cannot be relied on to show us the way.

In 1850, French economist Frederic Bastiat wrote his famous essay "That Which is Seen and That Which is Not Seen," which offers an important view of the classic libertarian economic thought and an important introduction to thinking about unintended consequences, both good and bad. Bastiat's argument is revived frequently to rebut the PP. In essence, the argument is that the use of the PP may allow society to avoid some impacts but may result in other hazards with worse consequences. The most effective of these arguments included Gokany's analysis of the use of dichlorodiphenyltrichloroethane (DDT) and GM food.[21] In essence, the argument acknowledges the impacts and risks of a decision but illustrates the costs or results of not using the technology. Gokany carefully constructs a case for the indoor use of DDT and for GM food using the PP. For example, he asserts that more human lives would be saved by indoor use of DDT in malaria-prone places and that such practices present little risk to ecosystems. Genetically modified foods would improve crop yields, extend ranges of food crops, and contribute to a reduction in world hunger. The benefits outweigh the risks in both cases, so precaution dictates that the technology be employed.[22] This argument is repeated in various forms many times.

There are several counterclaims made in response; for example, Ticknor argues "Precautionary action need not always mean banning a potentially hazardous activity. There must be ways to say 'yes'—with caution. Such tools and structures would shift the responsibility to those who create risks to examine and choose the most environmentally friendly options."[23] Among the components to which Ticknor refers to are the general duties to take preventive, precautionary action in the face of uncertainty that most reasonable people would acknowledge have always existed. Using this as a basis, government and businesses have a responsibility to act in a precautionary way if there is evidence that an activity (or substance) might pose a risk to health or the environment, even if there is no specific regulation of that activity.

Much of the criticism of the PP is rooted in the ambiguity of the terms used to describe it. Ticknor suggests that goal setting for environmental and public health protection would provide a stimulus for innovation and an acknowledgment of potential risks. For example, focusing on preventing harm and on methods of clean production has been demonstrated not only to produce benefits to the environment but also to increase economic competitiveness and innovation. Finally, Ticknor acknowledges that decisions made under a precautionary framework must be followed by continuous monitoring to ensure that they can be updated as new information becomes available. The environmental and public health goals should also be revisited to encourage continuous improvement and to avoid the unintended consequence.

Another response to Gokany's view is the value system employed. For example, Gokany argues that any single human life is worth more than any animal and "nonmortal threats to human life should take precedence

The Precautionary Principle and Design

over threats to the environment..."[24] Although he allows that there might be exceptions to the latter criteria, he does not provide a basis for making such exceptions. His view, however, does not recognize that most environmental impacts have at least some potential for human health impacts. Further decisions are rarely based on the life of a single animal versus a single human life; is a human life worth an ecosystem or a species? Would his argument extend to entire species and by default then to the whole ecosystem? The fact of the matter is that the basis for the PP is our inability to anticipate unintended consequence. Those who defend the traditional approach have not provided protection from unintended consequences in the past and often resist even acknowledging them, until forced to do so.

Gokany predicts that GM food will provide food for hungry millions and that, on its face, this is a justification for allowing it in spite of risks. History has shown us, however, that advances alone are not sufficient to feed the world's hungry. Distribution systems, public policy in support of high-end processed foods, and other aspects of food have as much influence as merely the quantity of food grown. Decisions being made regarding GM food are being made inside for-profit companies presumably for the benefit of shareholders. It is counterintuitive to expect a corporation engaged in economic competition to embrace the necessary requirements of public disclosure or shoulder the burden of precaution, without being compelled to do so. There is no reliable tradition of corporate altruism to rely on for how such technology will be employed and for whose benefit. Hunger in the world today is a reflection of political and distribution problems, not simply of supply. Further, what are the risks of GM food? How is the balance between feeding people and weighing those risks to be measured? In the end, how does Gokany measure future impacts in the interest of current satisfaction? Under the traditional system, materials, processes, and products were assumed to be safe, until proven otherwise. The burden of proof nearly always fell on government or third parties. Further, experience has shown that industries affected by demands for precaution engage in several tactics to create a false sense of uncertainty:

- Concealment of what they themselves know
- Hiring scientists inside and outside the corporation to both create "contrary" opinions and promote the notion of uncertainty
- Public relations and political "spins" to create the sense of uncertainty[25]

The impact is to delay action by public uncertainty. The impact of this sort of uncertainty has been decades of delay in responsive action to toxic exposure issues, ranging from lead, tobacco, asbestos, climate change, and alternative energy.

It is precisely because of our experience with uncertainty and unintended consequence that encourages the adoption of the PP. Sanford suggests that the interests of society and corporations could be served by identifying "precautionary matters." By listing such matters, society, through its civic infrastructure and regulatory process, would identify types of risks that would trigger the PP. In support of this, society should also establish legally binding principles that guide public and private institutions, acting in the absence of specific societal standards targeting the activities, to know when facts before them, or activities they are engaged in, may reflect a "precautionary" matter. Such a system would require an affirmative duty to study risks and alternatives and an obligation to maintain an open and transparent process. To protect individual rights, a prescribed preemptive privilege would apply (attorney/client, trade secrets, etc.) but be narrowly construed with regard to issues of public health and safety. Although not mentioned in the literature, the "unintended consequent" argument takes a very paternalistic view of the world and technology, in which the scientists and owners of technology reserve the right to decide for themselves how and what technology is to be employed.

For many, the fatal weakness of the PP is its attempt "to convert moral choices into legislation."[26] They feel that the principle oversimplifies very complex decisions. In the face of such complexity, the precautionary approach would require us to stop all research and progress because of uncertainty. Others argue, however, that the PP does not require research to stop at the outset. It only requires that, at each stage in the process, it be applied to determine if the anticipated benefits of going further outweigh the risks.[27]

An important criticism is that the PP does not provide a formal decision-making or analytic process. The form of the PP is such that it is unable to make specific decisions or conclusions; rather, it functions as a process that enables people to come to a conclusion. Attempting to rebut some of the criticisms of the principle, Ho writes "that the precautionary principle is not an algorithm for making decisions...but it is a principle on which to base decisions."[28]

Foster et al. observe that there is a precautionary continuum that ranges from weak to strong formulations. It is this variability that weakens the principle in the scientific community.[29] They make a science-based argument for the PP based on five guidelines:

1. Proportionality: Measures must be proportional to the desired level of protection and must not aim at zero risk.
2. Nondiscrimination: Comparable situations should be treated in similar ways, unless there is a compelling basis for not doing so.
3. Consistency: Measures should be comparable in nature and scope with what is known scientifically.
4. Costs and benefits of nonaction should also be considered.
5. Measures should be provisional, so that reassessment can be done as new data is available.

These guidelines provide important guidance and diffuse some of the more common criticisms of the principle and describe a reasonable process for continued progress.

The majority of arguments against the PP rely on applying the strong PP argument, which no one seriously proposes, or on promises of future good or benefits expected to accrue from technology that will more than offset risks or costs. Most arguments against the PP offer no reasonable alternatives, except the status quo. Only few critics deal with the difficult issues of translating precaution into a practical method of guiding public policy. Several items emerge from a review of both sides of the argument:

1. Even the weak PP presents a significant challenge to public policy makers and business.
2. Employing the PP will require an open stakeholder-oriented process.
3. Science has an important role to play in implementing the PP.
4. Precautionary decisions are forward looking and tend to be value based.

It is this last finding that leads us back to the purpose of this book.

Precautionary Principle and Design

It has been suggested that what is needed is a new approach to science. Unlike the image of the lone, self-motivated designers, modern design professionals are engaged in a competitive economic activity from within businesses. To the extent this is true, it must be asserted that there is an implicit burden on individual designers, scientists, or technologists to avoid harming others to the degree that they can reasonably anticipate the effect of their acts.[30] There is a basis to be found for this in public policy, for example, pesticide manufacturers and manufacturers of toxic materials have specific legal burdens in the United States to demonstrate the health and effects and environmental persistence of their products before they bring them to market.[31] Likewise, it can be argued that the moral reasoning exists for a precautionary approach on the part of designers of all stripes: engineers, architects, scientists, and technologists. There is a need for science and technology to consider the public good as well as the economic reward for their work. Concern for the safety and health of the public is a cornerstone of professional ethics for every design profession; indeed, it is this assurance that encourages the public to sanction the design professions.

Members of the design professions are subject to the parameters of the same social contract and moral minimums as any other individual in our society.

They act as moral agents in their daily lives and in their professional roles, but it should be argued that they have a higher threshold of responsibility by virtue and in the context of their professional expertise. Their special knowledge, in fact, allows them a greater opportunity to evaluate and understand the implications of their acts. The circumstance of that special knowledge must also include negative injunctions to avoid harm, but it may not require affirmative duties to do good as individuals. Businesses composed of design professionals may, however, have such affirmative duties similar to those ascribed to corporations by Simon et al.[32] (Chapter 4).

The concept of the moral minimum is described by Simon et al. as an "obligation not to impose social injury."[33] As a result of their special knowledge, the moral minimum for the competent professional must be a more focused threshold. Perhaps, the fabric of the social contract itself is changed for those possessing special knowledge. By virtue of their own codes of ethics and the practice laws of states, design professionals have some social responsibility to avoid harming others, and by extension that, a duty exists to correct social injury insofar as it is within their scope and authority.

Using the Kew Gardens principle[34] described by Simon et al., the moral minimum standard of behavior would include a conditional obligation to respond to a person needing help. Fundamental appeals to the moral basis for the PP could also be made on the basis of theories of justice. For utilitarians, what circumstance or combinations of circumstances would lead to the greatest satisfaction, happiness, etc. For Kantians, what behavior would a person want to universalize and apply in every situation? Rawls would require us to consider the behavior form the original position. In the end, it seems that the greatest satisfaction would result from an approach that made a reasonable evaluation of impacts before an act. It seems also that such an approach would meet the test of the Categorical Imperative and survive the deliberations from within the veil of ignorance.

The PP is not without problems in practice. It is difficult to anticipate the future effects of new technology and science, and such a careful approach may stifle innovation. The risks of the principle must be balanced against its alternative. If the actor does not capture the externalities of an act, then the risks associated with the act are spread among the stakeholders, but corporations and stockholders retain the benefits. In the past, corporations were able to benefit from new science and technology, often by avoiding costs of pollution control or other externalities. The results were that environmental quality was lost, public goods such as clean air and water were diminished, human health was unfavorably affected, and the quality of life was marginalized. In essence, the PP is the stakeholder theory applied to science, product design, and development. Equity, justice, and moral fairness require that the stakeholders share in the decision regarding what is reasonable risk. The burden to observe such precaution must lie with the originator.

In earlier chapters, the arguments for the ethical challenge presented to the design professions by sustainability was discussed. Those same arguments are made in the case of the PP. The constraints on individual design professionals working within organizations are of concern. The question must always remain: should a designer be ethically accountable for his own work? The answer must always be yes. It is the duty of the professionals to protect the public health and safety, to avoid causing harm, and to present harm when they can. The affirmative duty exists without regard for the circumstance. The duty, however, is not limitless. Within the stakeholder process, the designer finds the avenue to present her concerns, to have the implications weighed by those that might be affected, and to act accordingly.

The limits of professional obligation are described by the weak PP: Weak precaution states that full certainty that no harm will occur is not required to proceed with an action. Reasonableness is the measure, and reasonableness is the traditional measure of conduct and performance for design professionals. In practice then, the design professional should be expected to use the same precaution in his work. Sustainability requires only that the scope of concern be wider. The definition of public health and safety must expand to consider the environment and society.

What Do We Owe the Future?

Precaution, like design, is by nature a forward-looking act. As a decision-making principle, it is concerned with preventing harm or promoting the most beneficial outcome. In its most common use, the PP refers to protection of the environment and sustainable development. Among the most fundamental principles of sustainability is the idea of intergenerational equity. This is broadly defined as meaning the interests of the future must be considered in the decisions made by current generations. The PP rests firmly on the idea of intergenerational equity. The first consideration of the moral basis of the PP is to determine whether we do, in fact, have a moral duty to future generations.

The argument for a moral duty of the future has been approached from several different perspectives. First, however, is the assumption that there will be a next generation and one after that and so on. These future generations are, however, unborn. They do not consist of individuals that currently exist. Although such persons will have rights when they are born; nonexistent persons cannot have rights. Or can they? A right is an entitlement to act or have others act in a certain way. If you have a right to do something, then others have a correlative duty to act in a certain way. Human rights are universal, equal, not transferable, and cannot be relinquished. They are natural rights in that they are not derived from human institutions.[35] Negative human rights reflect the interests that human beings have from being free from outside interference, whereas positive human rights are interest that we have in receiving certain benefits (right to free speech).

Rights would seem to require a subject to which they can be ascribed. In the absence of a subject, there can be no rights or obligations.[36] Discussions of moral rights have the effect of moving the focus of moral judgment from a broad social context as a whole to the individual. Although some have argued that there is no temporal dimension to human rights, this is a difficult stretch: how do I consider my rights and obligations relative to generations past or future? Or they to me? There are no future or past persons relative to my rights or obligations. Further, should the future possess such rights, it would be reasonable that such moral obligations would be considered prima facie obligations.[37] As such, it would be an obligation easily overridden by more important obligations to current persons.

Instead, it is reasonable that unborn persons would acquire such rights at their birth and therefore could be described as having an interest in current actions that will affect them. Still, some suggest that the future persons have no right to complain, since it is up to us to determine whether they exist at all. This is the so-called futurity problem: "Our long term environmental policies cannot violate the rights of future generations, because the existence of particular future people depends on what we do. Unless they would prefer nonexistence, future people are in no position to complain that we behaved badly."[38] It seems to me that this principle has little substantive moral basis.

Therefore, to say that future persons have no current rights is not the same as saying that we have no obligation to take their interests into account. It is a challenge to attempt to predict how the future might value the environment, technology, or the world that they inherit. In general, such efforts do not fare well. Still, it is reasonable to assume that the value of clean air and water, biological diversity, available resources, and the like will continue to be important. Likewise, an environment without toxins, and other imposed risks, would seem preferable to the alternative. To the extent these are reasonable assumptions, we can identify the interests of the future and our correlative obligations.

Duty to Prevent Harm

Moral laws do not exist in nature. Hobbes posited that, in the state of nature, there is no immorality, nothing is prohibited, and nothing is required. In this view, laws of conduct come into existence only through agreement of individuals. Since there are no moral laws in a state of nature, every person has the right to do whatever is necessary for his or her own survival. Individuals leave a state of nature by voluntarily renouncing the right of nature and, in their own self-interest, accept the authority of a sovereign. According to Hobbes, then, the laws of nature for survival with other people might be, "Do not that to another, which thou wouldest not have done to thyself."[39] In coming to a society of people and leaving a state of nature, individuals agree to abide by the laws of nature and to the rules of conduct established by the sovereign. Hobbes suggests that there are two reasons why individuals must do this: (1) having made a promise to do so, they have a moral obligation to fulfil their promise, and (2) to not do so would be irrational because it would be acting against their self-interest.

John Stuart Mill expanded on Hobbes's thinking by stating that society has a duty to interfere with the actions of people to prevent the strong from preying on the weak, in short to prevent harm. He was concerned that social or government interference in our lives would result in "mindless conformity"[40] but held that society did have a duty to interfere to prevent harm. This maxim is known as Mill's Harm Principle, and it limits interference on personal freedom to what is necessary to prevent harm to others.[41] There is general agreement that we have a negative duty to not cause harm, and a positive duty to prevent harm if we are reasonably able to do so. It is reasonable to extend these duties to our obligations to consider the interests of future generations in our decision-making.

There is disparity, however, in how to assess harm. Act utilitarianism considers an act right if and only if the good or benefit of an act outweighs the bad or cost of the act. Good and bad are not determined solely in the terms of the agent but for the benefit and harm of all. On its face, this would seem to be consistent with the PP. Mill argued that rightness and wrongness are relative and occur as matter of degree: an act is right in proportion to its tendency to promote happiness. The more happiness it will produce, the more right it is. Act utilitarianism, however, might require a person to forego his or her own interests or happiness in favor of increasing the total happiness. Also, act utilitarianism ignores duties, justice, and the distribution of happiness; as long as the greatest utility or happiness is achieved, the distribution is important.[42]

Rule utilitarianism, on the other hand, distinguishes between an act and a rule. Rule utilitarianism requires that moral rules are observed and that we select these rules to follow based on the greatest utility or tendency to promote happiness. Mill suggested that we should follow moral rules and consider the utility of a choice only when there is a conflict between rules. According to these views, our moral objective is to produce the greatest total of goodness, no matter who will enjoy it. Utilitarianism is extremely demanding; we are always bound by an obligation to be improving the total happiness or utility, whether we benefit or not. Utilitarianism is also not concerned with distribution per se, only the total of happiness or utility. This approach would support a PP approach only insofar as the benefits or utility were increased for someone in the present, not necessarily in the future. It may also require current generations to sacrifice their happiness in favor of greater happiness in the future.[43] Where our duty requires us to consider the interests of the future, it does not make us subject to them. Hurka finds, "What is required of each generation is just that is pass on a total package of opportunities that is comparable to its own, whatever the exact composition of this package may be."[44]

Kant believed that all rational beings will pursue their own happiness and that when people act for a reason, that is, they decide to act, they are following a rule of action. From this, he derived the Categorical Imperative: "Act only according to that maxim by which you can at the same time will that it should become a universal law."[45] A universal law is understandable and can be discovered through reason, and it would apply to all people all of the time. He summarized the duties that were derived from the Categorical Imperative as: (1) everyone has a duty to help those in need, (2) gratitude because of a benefit is a duty, and (3) general respect should be given to other rational agents. At least the first and third duties speak directly to the PP.

John Rawls, on the other hand, believes that the correct standards of justice are those that rational, self-interested individuals in a certain hypothetical state analogous to the state of nature, which he calls the original position, would accept, if given the choice. It is important that the acceptable moral rules are those that everyone can agree on and that can be publicly justified. Further, in the hypothetical condition, Rawls requires that there be a veil of ignorance over the members of society, that is, they can have no personal information of themselves. All decisions about principles are made behind a veil of ignorance that ensures impartiality and objectivity[46] (see Chapter 2). It would seem reasonable to assume that, under such an approach, a reasonable person would choose a precautionary approach to decision-making to be sure that his or her interests were considered. It places an equal burden on all current generations to consider the future, as they would want themselves to be considered.

There is a strong moral argument for the PP, particularly when considering Kant and Rawls. Much of the world has started to embrace the principles of precaution; however, there is still wide disagreement on how to practically apply it. As a practical tool, it still lacks the formulation required. There is, however, significant resistance toward adoption of the PP in the United States. At the core of the resistance to the PP in the United States and elsewhere, there appears to be a lack of willingness to share decision-making and an institutionalized resistance to the necessary transparency. There are legitimate concerns with its use, but a lively debate should be a part of practice of sustainable design. The debate has been robust in some technical and professional circles but needs to be brought to the market and discussed more widely.

Precaution and Design

The application of the PP to the work of the design professions has likewise received little attention. Do designers have a precautionary obligation and, if so, to whom? Is the obligation to the client, to the end user, to society at large, or to all stakeholders? If precaution is morally justified, as argued earlier, then professionals do have a precautionary obligation to prevent harm, as they can reasonably do so through design. As noted earlier, the balancing of costs and benefits can be tricky business, however, and the design professionals must rely on their training and experience as well as the input from stakeholders.

Concern with the PP stifling or ending innovation, new products, or projects is justifiable only in the face of the strong PPs, just as it would be under strong sustainability principles. In essence, under the strong side principles, precaution and the avoidance of risk could become ironclad rules under which the unachievable test of certainty is applied to every action. Such a test is unreasonable, because it cannot be met and is absurd because human nature is driven to progress, to build, and to change. People will not abide the ridiculous constraint on progress for long.

The precautionary obligation of the design professions requires only that reasonable precaution and care are taken. Weaker, though not the weakest, precaution and sustainability rules should be followed. Stakeholder processes, consilience, and our awareness of the demand for sustainable design require and enable a broader definition of professional obligations. In a paradigm of sustainability, the designer must be educator, expert, consensus builder, the synthesizer of solutions, and the servant of the process. The PP provides a pathway for the designer to pursue sustainable outcomes. By diligently pursuing a precautionary approach, the designer demonstrates a reasonable effort to achieve sustainable outcomes in his work.

The designer's work is about bringing intentions to reality. It is an intellectual process at its root and cannot, in itself, be sustainable or unsustainable. Design is first ideas. The designer's job is to bring ideas to a form that can be translated, usually by others, into real products or outcomes. Bringing precaution to the process of ideation is a way of operationalizing the values that produce sustainable outcomes. Precaution also provides some transparency to the design process, which allows others to provide input and to evaluate the design as it takes shape.

The PP recognizes the uncertainties present in the designer's facts. We understand that we rely on science that may never be complete and that decisions relying on science will reflect what is known to be true at the time the decision is made. As the science changes over time, so will design decisions. The PP allows the designer some satisfaction that she is making the best decisions possible, given the knowledge at that time, and that should reflect a standard of care acceptable to any reasonable person.

Endnotes

[1] Gonzalez, Cathy, "The role of blended learning in the world of technology," http://www.unt.edu/benchmarks/archives/2004/september04/eis.htm, July 2009.

[2] Morris, Julian, "Defining the precautionary principle," in *Rethinking Risk and The Precautionary Principle*, Julian Morris, Butterworth Heinemann, editors, Oxford, UK, 2000, pp. 1–5.

3. Montague, P., "The precautionary principle," in *Rachel's Environmental and Health Weekly*, No. 586, 1998. http:monitor.net/rachel/r586.html
4. Morris, ibid, p. 6.
5. Calow, Peter, "The precautionary principle and what it means for your business."
6. Tickner, Joel, "Precautionary principle," in The networker, the newsletter of the science and environmental health net, May, 1997, Vol. 2, #4.
7. Boehmer Christiansen, Sonja, "Definitions of the precautionary principle," in *Interpreting The Precautionary Principle*, Tim O'Riordan and James Cameron, editors, Earthscan Publications, London, UK, 1994.
8. ibid.
9. ibid.
10. ibid.
11. Appell, David, "The new uncertainty principle," *Scientific American*, 284(1), 18–19, 2001.
12. Brombacher, Marcus W. "The precautionary principle threatens to replace science," *Pollution Engineering*, Summer 1999, p. 32.
13. Blanchfield, Ralph, "Commentary on the precautionary principle," February 6, 2000, http://www.biotech-info.net/PP_commentary.html
14. Cohen, Bonner R., "The perils of the precautionary principle," Issue Brief, Lexington Institute, http://www.lexingtoninstitute.org/environmental/precautionary.html
15. Tantillo, Jim, "The precautionary principle—A race to the bottom?" http://www.mailbase.ac.uk/lists/enviroethics/2000-05/0025.html
16. ibid, p. 4.
17. Appell, David, "The new uncertainty principle," *Scientific American*, 284(1), 18–19, 2001.
18. Levidow, Les, *"Sound science as ideology"* Centre for Technology Strategy, Open University, UK, levidow@open.ac.uk
19. Cranor, Carl, F., "What could precautionary science be? Research for early warnings and a better future" Dept. of Philosophy, University of California, Riverside, CA.
20. McAfee, Kathleen, "Social and ecological dimensions of precaution," Environmental studies Dept. University of California, Santa Cruz, CA. knmcafee@cats.ucsu.edu
21. Gokany, Indur M., *The Precautionary Principle; A Critical Appraisal of Environmental Risk Assessment*, CATO Institute, Washington, DC, 2001.
22. In fairness to Gokany in his conclusion, consistent with Bastiat, he states he seeks only that the precautionary principle consider all sides to an issue and that it be objectively applied. Others using Bastiat's argument however are less open.
23. Ticknor, Joel, "Precautionary principle: Current status and implementation," *Synthesis/Regeneration*, 23, 2000.
24. Gokany, ibid. p. 9.
25. Lewis, Sanford, "The precautionary principle and corporate disclosure."
26. Holm, Soren and John Harris, "Precautionary principle stifles discovery," *Nature* 400(398), 1999.
27. Howard, C. Vyvyan and Peter T. Saunders, "Sensible precautions make good science..." *Nature* 401(207), 1999.

[28] Ho, Mae Wan, "The precautionary principle is coherent" ISIS paper, October 31, 2000, http://www.biotech-info.net/PP_coherent.html
[29] Foster, Kenneth R., Paolo Vecchia, and Michael Repacholi, "Science and the precautionary principle," *Science*, 288(5468), 979–981, 2000.
[30] ibid.
[31] US Federal law, The toxic Substance Control Act.
[32] Simon, John G., Charles Powers, and Jon P. Gunnemann, "The responsibilities of corporations and their owners" in *Ethical Theory and Business*, 5th edition, Tom L. Beauchamp and Norman E. Bowie, editors, Prentice Hall, Upper Saddle River, NJ, 1997, p. 61.
[33] ibid.
[34] Based on the Kitty Genovese murder which occurred in the Kew Garden section of New York City, ibid, p. 63.
[35] Shaw, William H. and Vincent Barry, *Moral Philosophy and Business*, 8th edition, Wadsworth, Belmont, CA, 2001, p. 73.
[36] Beckerman, Wilfred, "The precautionary principle and our obligations to future generations," in *Rethinking Risk and the Precautionary Principle*, Julian Morris, Butterworth Heinemann, editors, Oxford, UK, 2000, pp. 46–59.
[37] ibid, pp. 70–71.
[38] Wenz, Peter, *Environmental Ethics Today*, Oxford University Press, London, 2001, p. 44.
[39] Hobbes, Thomas, Leviathan, 1651 as quoted in *Moral Philosophy, Theories and Issues*, 2nd edition, Emmett Barcalow, editor, Wadsworth Publishing, Belmont, CA, 1998, pp. 172–175.
[40] ibid, p. 232.
[41] Dubber, Markus Foundational Texts in Modern Criminal Law, Oxford Scholarship Online. November 2014, http://www.oxfordscholarship.com/view/10.1093/acprof:oso/9780199673612.001.0001/acprof-9780199673612-chapter-9.
[42] ibid, p. 136.
[43] Hurka, Thomas, Sustainable development: What do we owe to future generations? A working paper sponsored by the Centre for Applied Ethics.
[44] ibid.
[45] Kant, Immanuel, as quoted in *Moral Philosophy, Theories and Issues*, Emmett Barcalow, 2nd edition, Wadsworth Publishing, Belmont, CA, 1998, p. 148.
[46] Rawls, John, *A Theory of Justice Revised Edition*, Belknap Press of Harvard University, Cambridge, MA, 1999, pp. 102–130.

7

Flourishing

With every mistake we must surely be learning.

George Harrison

Throughout this book, I have focused on the design professional as a person set apart from other people because of the special knowledge that he or she has through education and experience. This special knowledge is recognized by society, and a certain exclusivity is extended to members of the professions; only those who have been accepted into the profession by the profession itself may perform certain types of work. With this sanction, the individual professional accepts certain obligations, chief among them are to protect and advance the health, safety, and welfare of the public at large. All other duties are subordinate to this one. My thesis is, in essence, that concerns for the environment and sustainable outcomes are in the interest of the public and directly and indirectly affect the health, safety, and welfare of people. To make the argument that sustainable outcomes fall within that obligation, one must extend the duty to future people and future outcomes.

As described in Chapter 5, there is considerable uncertainty when we set out to predict the future. Nature is complex, and we frequently do not anticipate outcomes, direct and indirect, of our actions. Still, we have no choice but to act on the best information that we have to anticipate outcomes and to mitigate those we find to be undesirable. In an important sense that is what design is, and always has been, about. Designing for sustainability is extending the scope of the designer's concern to include the environmental and human welfare implications of our design and the effects it can be expected to have in the future. In the end, we will assess design in terms of contributions made to, in Ehrenfeld's terms, human flourishing.[1] As this is written, we are in the process of learning what that means in terms of day-to-day practice and adapting to new expectations in the marketplace. As noted in the first pages of this book, at the very heart of this adaptation is a value shift and a change in the underlying philosophy of design. Perhaps, we are in a time where the old Epimethean philosophy is being replaced by an emerging Promethean ethos.

The design professional once again has a critical role to play. Few in society are as well prepared to understand the implications and challenges of a sustainable design ethic. In some ways, few have more at stake as well. Changes to one's design practice and the values that are the foundation for that practice are not easily made. There may be the sense that radically changing the underlying values means a refutation of prior work. This references the paradigm shift described by Thomas Kuhn in "The Structure of Scientific Revolutions."[2] Although Kuhn limited his observations to science and scientific revolutions, the process he described has wider applications and is particularly apropos to this argument. Paradigms then are accepted means and methods of practice, be it scientific research or design. To be accepted into a profession, students learn the accepted methods and means to the satisfaction of the profession; in the past, they did it through experience, and today, they do it via education and examination. To a large extent, the standard-of-care concept already discussed is in effect the working model of the paradigm within which a designer works.

The standard of care, though, is deliberately without precise definition, because it is recognized that no definition could ever encompass the myriad of circumstances and anticipate the changes that occur within the scope of any design task. Still, the standard is dynamic and does provide for change through the "reasonableness" test. What is reasonable will be different as circumstances and knowledge change. In the case of scientific revolution, it is necessary that there be an acceptance of an entirely new explanation or understanding of the facts. Once this occurs, scientists can no longer rely on the old paradigm, the old explanation. While revolution suggests a rapid displacement of the old in favor of the new, the length of time is less important than the change itself. The shift is revolutionary, because science is serial in nature; that is, an understanding is based on a more fundamental understanding, which in turn relies on yet more fundamental understandings and so on. New explanations of what is known, therefore, may ripple through the collected knowledge of how nature works. Those who have invested in their careers, have become successful, and have developed reputations under the old paradigm may resist the new way of explaining the world for understandably human rather than scientific reasons. Kuhn observes that much of science is not engaged in searching for truly new ideas but is looking for ways to eliminate anomalies and strengthen what is already believed to be true. Once the new paradigm is accepted, however, it is a repudiation of the old paradigm. This too may have resonance for the fields of design. As the need for sustainable outcomes becomes accepted, the standard of care will shift to reflect that, the old approaches to design will fall away, and past practices will no longer rise to a professional standard.

The question remains whether designers will lead or follow in this revolution. Of course, the answer is that they will do both. It makes sense that designers would be among the leaders in understanding the causes of environmental decline and the implications of growing populations. The various design professions look with pride to visionary leaders, but it may be illustrative to consider the work of scientists who have contributed significantly to our understanding of the issues we face. For example, Amory Lovins, a physicist by training, and the Rocky Mountain Institute (RMI) have produced as many sustainable design ideas in so many areas of design as any one source.

The work of RMI teaches us any number of lessons, but chief among them is the importance of the design paradigm used to address problems. Likewise, Michael Braungart, a chemist, working with William McDonough, developed an approach to their work that is perhaps more valuable than any individual product. Looking closely at the values these and other innovators bring to design reveals the design paradigm that they employ.

Of course, there are many designers at work every day on these fronts, but we must only look to the marketplace to see that the overwhelming products of design are currently unsustainable. Individual products and projects are celebrated and written up; successes are showcased. If we are to judge the performance of design today based on the magazines of the various professions and the popular press, one might think that the sustainable design problem was solved. However, a quick survey of the marketplace tells us that this is not the case. While the science of the twenty-first century has established the current condition of the natural systems on Earth and the anticipated effects of continuing our practices of the past, much of the routine design of today is firmly rooted in the last century, some of it even in the century before that.

The design revolution will occur because of pressure from several fronts: clients, regulators, and designers themselves. The design professions, however, have a clear moral and ethical obligation to promote sustainable outcomes, because they are in the best position to know the facts and to propose solutions and because they have a paramount duty to protect health and safety and promote welfare. Moral behavior is a matter of acting on reason. We act as a result of deliberation and of reflection on what we know and the implications of our actions/inaction and outcomes. This moral reasoning introduces the concept of "ought," as in we ought to do this or ought not to do that.[3] Ethics could be defined as how we put that "ought to do" into practice. In light of what is known about human activities and environmental quality, and the prospect of more and more people in the coming decades, what is it design professionals ought to do? Is design so different from science that once a new paradigm emerges and gains acceptance it should serve as both a way to assess design going forward and a repudiation of the old way? Once the need for sustainability is acknowledged and accepted, is it not true that the old demonstrably unsustainable design paradigms are repudiated? Therefore, is each act, decision, specification, and design that is not directed to a sustainable outcome unethical?

What Ought We Do?

It has already been noted that design professionals are only one part of the design process, and though they have special knowledge and authority (by virtue of their license to practice), they are not, in most cases, in a position to dictate the design. They have a duty to lead the process and to adhere to the obligations of professional practice, but they are not the only or even the most powerful actor at the table. We have noted that there must be some limits to the design professional's obligation, but, just as certainly, those limits are dynamic and change as the marketplace and knowledge change. As the demand for sustainable outcomes increases, the design professions will go along in response to the demand. But what is their role in promoting the change?

A partial list of the direct implications of current community design was given in Chapter 1. The impacts include increased obesity, heart diseases, high blood pressure, and other health impacts attributed directly, at least in part, to the pattern of suburban development that became so common after World War II but that morphed into their most common form since the 1970s. Products on the market with known risks to human health range from cosmetics to food. Streets are routinely designed for the benefit of vehicles but with known significant risks to pedestrians and bike riders. Food is "engineered" to be inexpensive but unhealthy, sometimes even with claims to good health on the labels. Many products are constructed by workers earning very low wages and laboring in unhealthy conditions. All of these are reflections of our intent, our designs.

Often these designs are consistent with prevailing public policy. Is only meeting the benchmark set by public policy sufficient for the design professional to assume that his burden of protection from harm is met if he has compelling information to the contrary? It cannot be so; otherwise, the standard of care is simply the minimum required by law, in which case there is no professional measure of "reasonableness." In such a case, meeting legal minimums becomes the definition of reasonable. We would not rely on the judgment of the professional in such circumstances but on the applications of standards, policy, and law in lieu of thoughtful analysis. There can be no professional obligation beyond the minimum if this is the case; reasonableness would merely be compliance, and the standard of care advances only as fast and as far as regulations can change.

It seems clear that professional obligations to protect go beyond the minimum. Recall the robust and significant changes to the codes of ethics by the American Society of Civil Engineers (ASCE) and the American Society of Landscape Architects (ASLA); these speak to the recognition the professions have to lead in matters of sustainable outcomes. Many of the professions require civic involvement and/or service to the profession as part of the professional's obligations. This seems to be an important aspect of professional life, as it relates to the needed paradigm shift in design values and philosophy. Perhaps, the ethical obligation in times of change is for the professions to conduct public education and awareness and to promote the ideas and values that the new paradigm of sustainability requires.

Designer as Teacher

Professionals are empowered with special knowledge and education that enable them to anticipate the outcomes of thier work. Professionals also have obligations to keep their skills current, to maintain a working knowledge of the practices and methods of their profession, and to understand how new information should be accounted for in their work. This obligation to maintain skills and knowledge is a commitment to consilience and to folding new knowledge in other professions and areas of knowledge into their own. As previously discussed, knowledge becomes substantive and worthy of acting over time, but once it reaches such a point, the design paradigm must, in fact, change. The designer works in a marketplace of ideas that rise and fall; design practice must reflect that dynamism.

The difficulty is that the designer typically also works in a competitive marketplace. As discussed in Chapter 4, in the competitive marketplace, the designer may lead the design process, but she is only one of a number of parties involved. New ideas, new paradigms emerge on different fronts and are accepted among the parties to design at different rates of speed. The designer's duty to protect public health and safety and welfare remains paramount but more difficult to operationalize as new information emerges that influences our understanding of health effects, safety concerns, and the complex character of what contributes to welfare. Most of the design professions include some aspect of public service in their codes of ethics or professional canons. It may be that among the most important services that the professions can provide is to educate the public on how the design paradigm is changing.

It can be difficult to act as teacher and design leader within a given project. The nature of the design process is sometimes adversarial, as different voices struggle to be heard. The design may have difficulty advocating unfamiliar design considerations as an actor within the process. The voice of the professional speaking as an advocate for new design values may be clearer when not representing a client or a project. The design professional can serve the profession, his own interests, and the interests of all parties to the design process by providing education to the public at large that is not project specific. By doing so, the parties to a later design may be better prepared to accept the new design paradigm, and the designer is spared the burden of re-educating parties already engaged in a project.

The fact is that, despite a general concern with the environment, there is frequently a disconnect between a general concern and individual acts. The noted natural history writer John McPhee wrote that "it seems to me that if people aren't interested in the earth and how it works, they are disenfranchised."[4] A person with no experience in nature cannot appreciate it; therefore, the farther our experience is from nature or a life with nature, the less is our appreciation. The farther we are from physical contact with nature, the less we value natural systems and the insulating impact of progress. Personal experience is then translated into cultural bias, as incremental changes occur in the attitudes of individuals. While 76% of Americans consider themselves environmentalists, the average American spends only 3% of his time outdoors.[5] A study released by the Department of Natural Resources in Maryland indicated that Maryland would have to protect 70,000 more acres (an area roughly twice the size of Washington, DC) if residents were to enjoy the same access (availability) to forest, parks, and other public lands in the year 2010. In a survey completed by the Maryland Greenway Commission, 89.5% of Marylanders surveyed thought that conserving land is a good way to spend tax dollars, but 49% felt that state and local government were doing enough to preserve natural resources/open space in their community. Despite the significant self-identification of people with the environment, the gap between firsthand experience and an appreciation of the environment grows.

In studying environmental values, Kempton et. al found that environmental values of Americans are derived from three sources: religion (environmental protection because nature is God's work), human-centered values (environmental protection for utilitarian reasons), and biocentric values (environmental protection because of belief in the intrinsic value of living things). The study suggests that a technical knowledge of the environment is not required but either experience or a casual knowledge is.

My own study found that adult environmental values are significantly affected by the character and quality of experience in adolescence. An important lesson to be drawn from the studies and research completed to date would be that environmental ethics can be derived from a variety of sources but that those that entail some firsthand experience seem to be more compelling.[6] The design professions should provide and support efforts to increase public access to nature and the environment and to reinforce the nature experience, especially in children.

One "ought" that designers might consider that embracing is promoting awareness of nature and environmental systems in their own work. This means a deeper understanding of the underlying principles of design and how they relate to natural laws and systems. Much of design serves to illustrate that designers themselves are often not making connections between their work and the environment. Access to inexpensive energy and a mind-numbing number of materials with wide-ranging performance characteristics has allowed much design to become input intensive but frequently intellectually vacant. Buildings are built without regard to how they might function to move air, provide light, conserve energy, and finally promote human well-being. Problems are solved by using more energy, bigger compressors, larger heating systems, and more electric lights. Architecture is commonly reduced to mere volume, with interchangeable decorative elements. Site work invariably destroys landscape and environmental functions, without any attempt to mitigate the impacts. A quick analysis of a thoughtfully designed building of 200 years ago frequently reveals more thoughtful design with regard to how the building works than most buildings constructed in the last 50 years. Consumer products are commonly designed to be inexpensive to manufacture, and there is a tendency to rely on materials and process that result in significant embodied energy but without regard for the final disposition of the product.

Consumer consumption represents about 70% of the US economy and is often singled out as a significant environmental concern. Consumption is a concern largely because of the scale of impacts caused by more than 300 million people in the United States and nearly 7 billion worldwide. In the United States, we generate 1.1 tons of solid waste per person each year. While the amount of waste generated per person has remained fairly constant, the population continues to grow.[7] These numbers do not include hazardous waste (3.2 pounds per day per capita) or wastes that are landfilled or disposed of on the site where they are generated.[8] Consumption is not the problem in and of itself; it is natural to consume things. Our sustainability issues with consumption lie primarily in resource depletion, energy intensity, and the generation of materials that are unsuitable for recycling or reuse. These are design issues. By bringing new values to the design paradigm, we will bring changes to these sorts of outcomes.

Design Values

In discussions with other design professionals, the issue of sustainability often involves trying to define what the word means. In Chapter 1, some definitions of sustainability were discussed in broad strokes, and ultimately, I brought the issue down to a question of values. Design proceeds after the designer's philosophy, and that philosophy is a reflection of the values of the designer, the client, and the culture that creates a context for the design. Current effort to produce sustainable outcomes reflects the values of the designers and society.

While this book is not limited to issues of architecture, the example of "green architecture" can be illustrative for design in general. Green building and operations provide the opportunity to employ proven design to reduce the energy demand and improve the performance of the built environment. Two thirds of the US electrical energy demand is from building operation of heat, air conditioning, and light. Energy efficiency can be improved through the use of natural light, building orientation, and material selection. Green buildings are found to be healthier and more productive spaces than traditional construction. The increased natural light is among the commonly appreciated features found in green buildings. Careful selection of materials such as floor and wall coverings and the design of a building to passively move air can contribute to cleaner indoor air. As global energy supplies tighten because of increased demand, and pressure mounts to respond to global climate change, green buildings are among the positive steps that institutions and business can take.

The costs of green buildings are typically no different or only nominally higher, whereas the benefits in terms of life cycle savings and operation costs are significant. Lower energy costs, water use, and waste generation are the most common areas of savings. Green buildings are healthier and constitute a more desirable working and living space. Green development encompasses a range of design, construction, and operational approaches that are employed to reduce the environmental impacts of development as well as to promote human health, well-being, and productivity. These approaches combine in ways that are variously called sustainable design, green architecture, restorative design, eco-design, and so on. Common elements shared by all of these are as follows:

- Use of regionally available, sustainably harvested, low impact, and reused/recycled materials
- Design sensitivity to energy use, the incorporation of energy conservation, and even renewable energy generation
- Water conservation
- Waste minimization
- The use of design to incorporate natural light, appropriate ventilation, and human scales
- Design and construction techniques that evaluate and minimize environmental impacts
- The use of site development practices to minimize and mitigate impacts

The impacts of development and urbanization are well understood. It has been observed that more than 80% of all buildings in the United States have been built since 1960. As our appreciation of the serious environmental effects associated with greenhouse gas emissions has advanced, the role of buildings and development has been more closely scrutinized. For example, the Department of Energy has found that 82% of the human caused greenhouse emissions are energy-related carbon emissions, and, of that, 48% of the increase in US emissions since 1990 is attributable to increasing emissions from the building sector.[9] In fact, buildings (residential and commercial) are responsible for more greenhouse gas emissions than either the industrial or transportation sectors. As the US population continues to grow, the construction of more buildings is anticipated. In the United States,

50% of the electricity generated is from burning coal, the least efficient and the dirtiest of the major fuel sources.

There is a significant variety to green buildings. Some buildings are considered to be "high performing" because of energy efficiencies, though generally, these buildings are not considered green in a strict sense. Green buildings exist on a continuum ranging from fairly straightforward attempts to reduce the environmental footprint of a building to very sophisticated structures. There is not a "one-size-fits-all" formula. Several guidelines have emerged to guide and measure the process of designing, building, and operating a green building. The National Institute of Standards and Technology (NIST) has developed the Building for Economic and Environmental Sustainability (BEES) model with support from the Environmental Protection Agency (EPA) and Housing and Urban Development (HUD). In essence, this approach evaluates building materials for their life cycle costs and environmental impacts in 10 areas of concern. Designers or builders are able to determine the environmental loading of a material before they specify its use.

The American Society for Testing and Materials (ASTM) has developed an approach for evaluating the life cycle costs of building materials (E 1991 Standard Guide for Environmental Life Cycle Assessment of Building Materials/Products). This approach is broad based and includes consideration of embodied energy, raw materials acquisition, and environmental impacts. The best-known approach is the US Green Building Council's Leadership in Energy and Environmental Design (LEED) program. This approach provides design professionals, builders, building owners, and stakeholders to consider the various ways in which building design and operation impact environmental performance. LEED uses a point-based system to guide choices that range from building location and orientation to building materials, landscaping, water and energy conservation, light pollution, and even transportation. Eligible buildings may apply for and be recognized as certified or silver, gold, or platinum certified, depending on the number of points they achieve.

It is believed that this comprehensive approach is chief among the reasons why LEED-certified buildings have been shown to have significantly lower operating costs and healthier and more productive occupants, in addition to protecting the environment. Unlike BEES and the ASTM method, LEED does not certify materials or processes, though these approaches might be used by designers specifying for a LEED project. LEED-certified professionals are employed in the planning, design, and construction processes. There are specific guidelines published for each area, and a Green Building Rating System describes how design choices will be rated in terms of LEED points.

The data on cost implications of green buildings is difficult to synthesize to any specific building project. It is more cost-effective to build green ideas into a new building or site, although retrofitting offers advantages in some cases. Cost comparisons are difficult to assess because true or life cycle costs of traditional construction are rarely accounted for. Green materials tend to be more durable than traditional construction. This may result in higher construction or development costs, but value is returned in much longer life cycle and maintenance costs. Green buildings have been shown to result in higher productivity than traditional construction. In general, green construction has been found to be cost competitive as capital projects and more cost-efficient from operations standpoints. Figures routinely used on the literature without supporting documentation suggest that green architecture increases construction costs from 2% to 11%,[10] although some studies suggest that the preconstruction cost estimates of green buildings are often overstated by as much as three times the actual cost.

Among the most comprehensive studies of the cost of green buildings was the 2004 Davis Langdon study.[11] This study compared the costs of green architecture to traditional architecture from two perspectives: (1) what would be the cost difference between a traditional building and the same building as a green building, and (2) a direct comparison of 45 buildings seeking LEED certification and 93 buildings not seeking any "green" certification. The study looked at laboratories, libraries, and classroom buildings for these comparisons. In general, the study drew the following conclusions:

- Most green buildings can meet the goals of LEED within or nearly within the original program budget.
- Many non-LEED buildings included green elements for purposes of operational efficiency, without specific regard to environmental performance or certification.
- The key to the cost performance of green buildings lies in the program planning phase, the expertise of the project team, and project management/oversight.

The costs of green buildings have also been evaluated in a study commissioned by the US General Services Administration (GSA). The 2004 study evaluated the costs for buildings seeking certified, silver, and gold LEED ratings. The study only considered courthouses and building modernization but concluded that costs may range widely, depending on the rating being sought. In cases where low or no cost features are used, the study found "the overall cost premium [to be] surprisingly limited, even at higher rating levels. Under certain conditions it is possible to show a slight decrease overall."[12]

These and other studies demonstrate that, in general, the capital costs of green buildings are often no more or only nominally more than those of traditional buildings. Though the Davis Langdon study strongly suggests that cost performance can be related to the management of the process from planning through construction, other studies have indicated that, although costs may rise slightly, as greater efficiency is designed into buildings, the rate of savings increases more dramatically.

Operation cost data is highly variable, but in general, the following sorts of savings found in green buildings are most often associated with lower energy costs. Electricity costs are in the range of 45% to 55% less,[13] and in even more sophisticated energy conservation systems, the payback periods are estimated at 7 years to 9 years. More common energy efficiency feature has a payback of less than 2 years. A study completed in Minnesota that looks specifically at schools and other public buildings found that the median payback period for school buildings was 2.6 years and for libraries was 2.1 years.[14] This study found that high-performing buildings saved on average $0.87 per square foot per year in lower energy costs. A total of 3% savings were derived from more efficient lighting- and load-responsive heating, ventilating, and air conditioning (HVAC) equipment. Water use is reduced as much as 90%, although figures of around 50% are more common.

Among the chief advantages of green buildings are the excellent life cycle costs. It is estimated that the life cycle savings of the typical green building are 10 times the initial investment; that is, for every $100,000 in initial capital investment, a saving of $1,000,000 is achieved over the life of the building.[15] In addition to the capital and operational costs of building, there are benefits of green buildings that extend beyond the environmental performance of the building. Green buildings have been found to increase worker and student productivity. Worker productivity improvements ranging from 6% to 12% have been measured in green buildings.[16] Such significant improvements have been made in worker productivity in these high-performing buildings that it has been observed by some that companies can hardly afford to not build green! Similar improvements have been seen in worker turnover.

Performance improves in green schools as well. Studies have found that indoor air quality and classroom lighting have direct effects on both teacher and student performance. Studies have parsed the influence of various types of lighting and air quality on student performance. While green buildings commonly feature significant day lighting, studies show that the character of day lighting should be carefully considered and controlled. Likewise, the quality of air and acoustics within the classroom have a direct influence on student learning. While these studies focused on elementary and secondary schools, the results are notable. For example, classrooms with improvements in lighting alone were correlated with 7% to 18% higher scores in end-of-the-year examinations over classrooms without improvements. Students in the improved classrooms also progressed 20% faster in math exams and 26% faster in reading tests.

This brief overview of just one type of green or sustainable design illustrates the significant advantages and benefits, direct and indirect, of a more enlightened approach to design. Similar possibilities exist for all areas of design, and with the possibilities come opportunities for designers. Energy efficiency (a unit of energy per dollar of the Gross National Product) is a central concern of green development. Since the 1970s, the energy efficiency of buildings in the United States has remained flat, but the number of buildings has grown exponentially, so the environmental footprint of buildings has grown. Since the early 1990s, the European Union (EU) has sought to increase energy efficiency across the board, and in 2002, it passed rules requiring efficiency in commercial and government buildings. The contemporary office building in the EU is four times more energy efficient than its counterpart in the United States. In addition to the environmental advantages, the EU building is 75% less expensive to operate. Better buildings in the United States will achieve similar results.

So, what is the design philosophy at work here? And what are the values on which it is based? To define "values" is a difficult terrain to cover. We translate values into a working philosophy that differs in nuance and focus from one to another but that shares some common, albeit poorly defined, root values. Some values, such as beauty, are entirely subjective. Values such as fairness and equity might be defined by the same person in different ways under differing sets of circumstances. There is not a rigid, codified set of values that we can point to that are consistent from person to person. We can probably agree, however, that values such as beauty and justice are among the values we share, without having to pin down the details as to what they actually mean. Using the test of what a reasonable person might do to assess the standard of care is an acknowledgment that such a standard is beyond definition. Instead, "reasonableness" refers to an unspecified range of possible decisions or outcomes that would defy more precise definition. John Ehrenfeld writes in his book *Sustainability by Design* that a better term than sustainability might be "flourishing."[17] Sustainability in this context is the set of conditions that lead to or provide for us as people and individuals to flourish. What values could we envision that would underwrite such conditions? The LEED-certified buildings reflect values such as stewardship in the concern for energy and material efficiency and conservation but also values such as healthfulness and vitality, as evidenced by the natural light, careful attention to air flow, and change. Indirect results such as improved student learning and increases in productivity and employee retention suggest that these designs contribute to human welfare and flourishing.

If, indeed, flourishing is the result of the values that comprise the design philosophy, what are those values? If we are interested in sustainable outcomes of improved and preserved environmental functions, biodiversity, managed resource use, opportunity for human fulfillment, social equity, civic engagement, and respect for nature and for the future, what values will lead to those outcomes? At least a partial list of values comes to mind: healthfulness, justice/equity, and connectedness. These values are all extrinsic values in that they acquire their importance and desirability from actions not necessarily contained within them. They result from what we do. For example, healthfulness implies conditions that will lead to one being and staying healthy. Justice and equity or equal opportunity and fairness are all dependent on interactions, treating people as people Kant might say, rather than as the means to an end. Connectedness refers to access to other people, to information, to nature, and to the various institutions and traditions that make up society. To the extent that they are extrinsic, they are political and definable in terms of performance standards. We can create a standard that describes the human and environmental health concerns and outcomes we desire and measure design outcomes against that expectation. These extrinsic values extend beyond current interests and at least attempt to consider future interests as well.

Designer as Student

While it is not a value per se, it also seems as if designers must begin to take a longer view of their work. While design is generally an optimistic and future-oriented undertaking, we may need to define what the future actually is in terms of design. Is the future defined in terms of the life of the design object? Does it include the disposition of the materials used to create the object? Are externalities of the design considered in its life? Designers concerned with and working for human flourishing will necessarily take the long view of their work. What is an appropriate time frame for a design "horizon"? The time horizon for most designs seems to end at the point where the intended user takes over. The design life of the object is a matter of the design itself, so the designer's concerns end when the user begins to use the project. These truncated measures of the future are in terms of months, a few years, or perhaps a couple of decades.

How would our design change if we were asked to consider our design, its materials, and its implications in terms of the life span of a right whale (60 years), red maple (115 years), an American box turtle (123 years), a Galapagos tortoise (193 years), a ponderosa pine (up to 300 years), or a bristlecone pine (4600 years)? How would a longer view change design? Clearly, sustainable outcomes and human flourishing require a longer view by the designer. Material choices, for example, are a reflection of the time horizon selected for a project. Short-term products will result in the lower material efficiency and will return materials back into circulation sooner. Outdoor furniture that degrades in the presence of sunlight reflects a short time horizon in the choice of materials. What thought is given to the disposition of those materials when the product is no longer safe to use? Designs that require the destruction of valuable, irreplaceable top soil and agricultural land for construction of a project that has an anticipated effective life span of less than 50 years when urban sites are plentiful reflect a remarkable disconnect between natural systems, our place in those systems, and short-term satisfaction. When we specify a material, a sustainable design consideration would be to account for and plan for the disposition of that material at the end of its functional life. The designer's consideration cannot be limited to the time outlined in the construction timetable or manufacturing process.

Other values we might consider include beauty, mindfulness, and respect. These, of course, are intrinsic values that derive their initial value at the personal level, and they result from learning. These are what Aristotle might have called virtues. As discussed earlier, these will resist definition because of their subjective nature, but they serve an important underlying importance in creating desirable sustainable outcomes. Clearly, no meaningful lasting performance standard for beauty can be created, and likewise, for mindfulness or respect. The outcomes from designers practicing these virtues, however, would go some distance toward sustainable outcomes. Mindfulness of our impacts on our environment, and on each other, of how our design influences our healthfulness and that of the future are necessary elements of a truly sustainable design and future. Respect for natural systems and the implications of design for users and nonusers of the product or place would replace the hubris of design that has diminished the quality of our environment and that of the future.

An important part of the change that is underway in design is whether we will become sustainable, whether we are able to design a world where we can flourish, or whether we are merely trying to be less unsustainable and push the costs of our choices on to the future. In his book *Architectural Design and Ethics*, Thomas Fisher observed that few architects return to their finished work to learn what worked and what didn't.[18] As we attempt to learn how to design for sustainable outcomes and flourishing, some of the best lessons may be in our own previous work. Examining an old design with new eyes offers us the opportunity to learn, not from our mistakes but from the new design paradigm that we are learning to embrace. Old design reflects the values we employed at the time we did that work. Like scientific revolutions, as a new design paradigm emerges, it repudiates the old design paradigm. What was believed to be true is no longer so, and it requires of us to adapt to the new realities if we are to rise to the obligations professed by our professions.

In the end, Professor Dr. Anderson has it correct: the answer to the ecologic crisis is a question of philosophy. When he retells us of Zeus sending Hermes to give people the civic arts "to bring respect and right among men, to the end that there should be regulation of cities and friendly ties to draw them together." Hermes inquires if these civic arts should be distributed as were the arts stolen by Prometheus (one given the medical arts, another carpentry, a third farming, and so on). Zeus says, no, "Let them all have their share; for cities cannot be formed if only a few have a share of these as of other arts."[19] So, it would be, in this telling, that designers of industrial product, bridges, communities, gardens, buildings, and so on should lead in the civic arts for the good their work will produce. They should lead because their special knowledge has prepared them to do so and their ethical standards require it. Good design has come to have a deeper meaning, and a human flourishing must become a design value if we are to be sustained.

Endnotes

1. Ehrenfeld, John R., "Sustainability by Design," Yale University Press, 2008.
2. Kuhn, Thomas S., *The Structure of Scientific Revolutions*, 3rd edition, The University of Chicago Press, Chicago, IL, 1996.
3. Rachels, James, "What would a satisfactory moral theory be like?" *Moral Issues in Business*, 8th edition, William H. Shaw and Vincent Barry, editors, Wadsworth Publishing, Belmont, CA, 2001, p. 90.
4. McPhee, John, "Basin and Range," *Words from the Land*, Stephen Trimble, editor, University of Nevada Press, Reno, Nevada 1995.
5. Kempton, Willett, James S. Boster, Jennifer A. Hartley, *Environmental Values in American Culture*, MIT Press, Cambridge, MA, 1995.
6. Russ, Thomas, "Landscape ecology and individual values," *Proceedings of the 1997 Annual Meeting of the Soil and Water Conservation Society*, Soil and Water Conservation Society, Iowa, 1997.
7. ASCE, Report Card for America's Infrastructure http://www.infrastructurereportcard.org/fact-sheet/solid-waste, July 2009.
8. Zero Waste America, Statistics, http://zerowasteamerica.org/Statistics.htm, July 2009.
9. Battles, Stephanie J. and Eugene M. Burns, "Trends in building-related energy and carbon emissions: Actual and alternate scenarios," presented at the Summer Study on Energy Efficiency in Buildings, August 21, 2000, http://www.eia.doe.gov/emeu/efficiency/aceee2000.html#carbon_trends, 11/2/2006.
10. Kats, Greg, Leon Alevantis, Adam Meran, Evan Mills, and Jeff Perlman, "The costs and financial benefits of green buildings," *A Report to California's sustainable Building Task Force* 134, 2003, http://www.ciwmb.ca.gov/greenbuilding/design/CostBenefit/Report.pdf
11. Davis Langdon Seah International, http://davislangdon-usa.com/publications.html October 2004, 11/17/06.
12. Steven winter Associates, General Services Administration "GSA LEED Cost Study, Final Report" October 2004 http://www.wbdg.org/ccb/GSAMAN/gsaleed.pdf
13. Fahet, Valerie, "Building green always made sense: Now its beginning to pay off," 9/11/2005. The Chronicle, www.sfgate.com/cgi-bin/article.dgi?file=/c/a/2005/09/11/REG4DEKFQD1.DTL&ty
14. The Weidt Group, Minnesota Office of environmental Assistance, "Top 6 Benefits of High Performance Buildings," June 2005, http://www.moea.state.mn.us/publications/highperformance-brochure.pdf
15. Kats, Greg, Leon Alevantis, Adam Meran, Evan Mills, and Jeff Perlman, "The costs and financial benefits of green buildings." *A Report to California's Sustainable Building Task Force*, 134, 2003, p. 7.
16. Heschong Mahone Group, Inc. *Windows and Offices: A Study of Office Worker Performance and the Indoor Environment*. California Energy Commission, Sacramento CA, 2003, www.h-m-g.com/projects/daylighting/summaries%20on%20daylighting.htm

[17] Ehrenrenfeld, John R., *Sustainability by Design*. Yale University Press, New Haven, CT, 2009.
[18] Fisher, Thomas, *Architectural Design and Ethics*, Architectural Press, Hudson, 2008.
[19] Anderson, Albert A., "Why prometheus suffers: Technology and the ecological crises," *Society for Philosophy and Technology*, 1, (1/2), 28–36, 1995.

Appendix A

A survey of landscape architects was conducted to determine the influence of changes made to the American Society of Landscape Architects on the introduction of sustainable practices to design. The survey concluded that most design professionals' decision to adopt sustainable practices were based primarily on external factors such as the wishes of clients and competitive pressure rather than ethical concerns. For most design professionals, knowledge of the relevant sections of the code of ethics was limited only to an awareness.

Background

In the early 1990s, the American Society of Civil Engineers (ASCE) and the American Society of Landscape Architects (ASLA) each amended their code of professional ethics to include robust statements regarding the protection of the environment and sustainable development practices. The revisions were such that positive and negative obligations would be required of members complying with the codes of ethics. The purpose of this project was to assess the impacts of the amendments in both organizations; in essence, did the changes have the deserved influence on the respective professional practices? The objective of this project was to assess the influence of changes made to the codes of ethics of ASCE and ASLA and, specifically, to evaluate the impact of these changes on the professional practice of its members.

Scope of Proposed Project

The project attempted to assess the effectiveness of the changes in the code of ethics for both ASCE and ASLA. The scope of the project included:

- An evaluation of the amendments approved by each organization, including the obligations these revisions required of members.
- A brief historical look at each organization to provide a context for the change.

- Actual interviews with members of each organization and appropriate staff.
- Preparation of a short survey and the collection of data. This element is subject to the identification of a reasonable method of distribution and collection of survey data and the receipt of a sufficient number of answers necessary to yield significant findings.
- Collection and analysis of interview materials and survey data.
- Preparation of a paper summarizing actual data collection, analysis, and conclusions.

Limitations of Project

The proposed study was not intended to be exhaustive, and several difficulties were encountered. While every reasonable effort was made to construct sampling instruments that were valid and would yield significant results, the following limitations were encountered:

- The resources of the student were limited in terms of access to members of the professional societies.
- Wide distribution of the survey instrument yielded surprisingly low response. Although surveys were distributed to randomly selected firms and individuals via direct email and through a list serve, only 91 responses were received. Nine of the returned surveys were deemed unusable because they were substantially incomplete.
- Engineers, in particular, elected not to participate. Though more than 30 civil engineering firms and more than 50 individuals were contacted, only five responses were received from civil engineers/ members of ASCE.

The effects of these limitations are important to note. The low response rate to the random sample approach eliminated the possibility of the survey instrument providing a scientifically defensible degree of significance. Further, the extremely poor response from civil engineers eliminated the possibility of including them for consideration. The survey instrument was designed to be a random sampling; however, insufficient total responses were collected to provide significant results.

On the other hand, the survey instrument was designed to provide for both pragmatic[1] and convergent validation,[2] so the confidence in the instrument could be demonstrated. Since civil engineers were eliminated from consideration by virtue of the low numbers of respondents, the data collected form landscape architects could be used as a judgmental sample. In the context of

Appendix A 163

the inquiry, it was felt that landscape architects represent a fairly homogeneous group. Although there are demographic differences between individual members of the landscape architecture profession and with ASLA, there is a high degree of homogeneity with regard to the narrow area of concern. Further, the responses showed a fairly high degree of consistency; still, the margin of error was about 10%. Based on this, the investigation proceeded despite the low number of responses and the fairly high margin of error.

Survey Analysis

A total of 75% of respondents were landscape architects, only 6% were civil engineers. Nineteen percent were nonprofessionals or not licensed. However, 92% were members of ASLA. Ninety-three percent were working in private practice. A total of 27% had 5 or fewer years in the field, 27% had from 5 to 10 years of experience, 13% had from 11 to 15 years, and less than 1% had more than 16 years of experience. The majority of respondents (86%) said that they worked in organizations with between 26 and 50 employees.

The majority of respondents said that they were aware of the changes to the code of ethics (82%), and 71% had read the revisions. When asked to evaluate their own knowledge of the code of ethics, the majority (62.5%) indicated their degree of knowledge to be "an awareness." Only 25% indicated that they have a working knowledge of the code of ethics. Similarly, when evaluating the professional in general, 65% said that there was only an awareness of the code of ethics, only 12% said that members of the profession enjoyed.

Most agreed that the change in the code had made an impact (66.7%) and that engineering/landscape architecture practice has changed because of the revisions (60%). Sixty-two percent felt their practice had changed, and 66% felt the change to be a positive one. When it came to assessing wider performance issues, only 40% indicated that they knew of recent projects that did not comply with the spirit or the letter of the revised code (53.3% said that they did not know of any). Sixty percent felt that the revised code of ethics does not require a professional to adhere to the use of sustainable practices. Responses were evenly split (47% yes and 40% no) that the revisions made the use of unsustainable practices unethical. There is a high degree of consistency in these findings that suggest some reliability despite the high margin of error. Every respondent (100%) agreed that the code does impose a positive duty for professional members to consider environmental protection and sustainability among the concerns for public health and safety.

Well within the margin of error, most agreed that the code of ethics requires a professional to report unethical behavior (80%) and that a professional is required to report a significant risk (93%) or hazard (87%) to public health and safety. Further, most believed that within the scope of professional activity, the professional has a duty to prevent harm (93%) and to avoid acts that may cause harm to the public (100%), but only 33% felt that there was a duty to report the use of unsustainable practices.

Concern for the environment was also fairly high: 93% felt that within the scope of professional activity, the professional has a positive duty to protect the environment, but 7% did not feel that way, and each indicated in some fashion that the question was too vague to answer. Eighty percent felt that negative environmental impacts or a loss of environmental quality is to be considered harmful.

Most respondents (93%) indicated that issues of sustainability were common considerations in their design practice. Forty-seven percent of the respondents estimated that 11% to 25% of their work would be considered "green" (7% indicated from 26% to 45%, 7% indicated less than 10%, less than 1% indicated 46%–60% and 61%–75%), and 50% indicated that issues of sustainability were considerations in their practice, even 10 years ago (1992). Seventy-eight percent indicated that the revisions to the code of ethics were not a factor in their concern with sustainability.

The three most significant obstacles to a broader appreciation of sustainable development practices were (1) client's concerns or direction, (2) local ordinances and practices, and (3) reviewers. The three most important factors in the application of sustainable design were (1) clients concern or direction, (2) firsthand experience, and (3) changes in the standards of professional practice.

Conclusion

There is a significant degree of awareness among landscape architects of the professional's obligations to public health and safety and to prevent harm. This awareness includes the consideration of environmental quality and sustainable practices, but there is less agreement as to whether the professional's obligation requires the use of sustainable practices, even though most agreed environmental impacts represented a form of harm. There was an even split among respondents as to whether the use of unsustainable practices would be considered unethical. It would appear from the survey results that there is a great deal of awareness among landscape architects of issues of sustainability, though a somewhat lower degree of awareness of the standard of care required by the ASLA code of ethics. Approximately half of those surveyed do not consider sustainability to be an issue of professional ethics and favor a fairly narrow view of the obligations required by the code of ethics. This would be consistent with the level of awareness of the code, since about half

Appendix A

of those surveyed indicated that they had only an awareness of the revisions to the ASLA code of ethics; still, a majority of those surveyed indicated that the code of ethics had something to do with their decision to include sustainability among the considerations of design.

It would appear that the most significant factors in the wider use of sustainable practices were not ethical. Instead, sustainability is being driven, according to the survey, by the interests of clients, the knowledge of the professional, and the change in what is perceived to be the professional standard of care. This indicates that the revision to the code of ethics may be more in the nature of form than of function; landscape architects appear to adopt sustainable design practices because of external factors rather than because of ethical reasons.

Survey

1. Are you a licensed design professional?

    Civil engineer 6.3%

    Landscape architect 75%

    Other 18.8%

2. Are you a member of ASCE or ASLA? ASCE 6%, ASLA 92%

3. Are you aware of the changes pertaining to the environment and sustainability made to the professional code of ethics? 82% Yes, 18% No

4. Have you read the revised code of ethics? 71% Yes, 24% No

5. Do you agree that this revision was appropriate? 84.6% Yes, 0% No (15.4% did not answer the question yes or no)

6. Would you say your knowledge of the revised code of ethics could be described as (check one):

    0% Excellent

    25% a working knowledge

    62.5% an awareness

    12% incomplete

    0% Poor

7. Has the practice of engineering/landscape architecture changed because of the revisions? 60% Yes, 40% No

8. Has the change in the code of professional ethics made an impact? 66.7% Yes, 33.3% No

9. In your opinion, would you say that the knowledge of the revised code of ethics within the profession in general would best be described as (check one):

12% Excellent

65% a working knowledge

5% an awareness

12% incomplete

0% poor

10. Has your practice changed in general? 62.5% Yes, 37.5% No
11. How has this change impacted your practice?

    14% negative impact on business, 77% positive impact on business
12. Do you know of recent projects that do not comply with the spirit of the code of ethics? 40% Yes, 53.3% No
13. Do you know of recent projects that in your opinion do not comply with the letter of the code of ethics? 40% Yes, 53.3% No
14. Of the following, which would you say are the most significant obstacles to a broader appreciation of sustainable development practices (please rank them in order of importance, 1 being the most important, 2 the next most important, etc.)

    The following matrix shows the percentage of respondents ranking each factor on a scale from 1 to 9.

Factor	1	2	3	4	5	6	7	8	9
Competitive pressure from other firms not complying	8	–		15	–	23	23	–	31
Public policy (national)	8	–	7	15	15	15	23	7	7
Public policy of regulations (state)	–	8	15	31	–	15	7	15	–
Local ordinances and regulatory practices	33	26	20	–	13	6	–	–	–
Reviewers	8	23	23	16	8	8	8	–	8
Client's concerns or directions	29	22	14	7	21	–	–	7	–
Inability to find qualified staff	–	8	–	8	8	8	8	8	46
Lack of awareness of changes in professional practice	8	15	–	8	31	15	15	–	7
Firsthand experience with sustainable practices	–	5	11	–	–	5	11	16	53
Other									

15. Which of the following would you say have been the most important factor in the application of sustainable practices:

    The following matrix shows the percentage of respondents ranking each factor on a scale from 1 to 9.

Appendix A

Factor	1	2	3	4	5	6	7	8	9
Competitive pressure from other firms not complying	14	14	–	–	–	14	–	29	29
Public policy (national)	20	–	10	–	10	20	20	20	–
Public policy of regulations (state)	–	12	–	25	25	13	–	25	–
Local ordinances and regulatory practices	38	13	25	–	12	–	12	–	–
Reviewers	–	13	13	37	–	12	12	15	–
Client's concerns or directions	63	25	–	–	10	–	–	–	–
Availability to find qualified staff	–	–	–	–	29	19	–	29	31
Change in standards of professional practice such as the revised code of ethics	–	–	45	11	11	11	11	–	11
Firsthand experience	–	44	22	11	–	–	11	–	11

16. In your opinion, does the change in the professional code of ethics require a professional member to adhere to sustainable practices? 40% Yes, 60% No

17. In your opinion, does the code of ethics include a positive duty for professional members to consider environmental protection and sustainability among the concerns for public safety and health? 100% Yes, 0 No

18. In your opinion and in light of the revisions, is the use of unsustainable practices unethical? 47% Yes, 47% No

19. Does the professional standard of care require a professional to report unethical behavior? 80% Yes, 13% No

20. Does the professional standard of care require a professional to report unsustainable practices? 33% Yes, 61% No

21. Does the professional standard of care require a professional to report a significant risk to public safety or health? 93% Yes, 7% No

22. Does the professional standard of care require a professional to report a hazard to public safety or health? 87% Yes, 13% No

23. Within the scope of professional activity, does the professional have a duty to prevent harm? 93% Yes, 7% No

24. Within the scope of professional activity, does the professional have a duty to avoid acts that may cause harm to the public? 100% Yes, 0 No

25. Within the scope of professional activity, does the professional have an obligation to protect the environment? 93% Yes, 0 No (other respondents did not answer; several noted that the question was too vague or ambiguous)

26. In general, are negative environmental impacts or a loss of environmental quality harmful? 80% Yes, 13% No
27. Are issues of sustainability common considerations in your design work? 93% yes, 6% No
28. What percentage of your design work would be considered "green"?

7% 0–10%	<1% 61–75%
47% 11–25%	_____ 75–90%
7% 26–45%	_____ 91–100%
<1% 46–60%	

29. Were issues of sustainability common considerations in your design work 10 years ago? 50% Yes, 29% No
30. If you answered yes to question 29, was the change in the code of ethics a factor in your concern with sustainable design? 7% Yes, 78% No
31. Is your organization public or private? 7% Public, 93% Private?
32. Number of employees organization-wide:

_____	0–10
7%	11–25
86%	26–50
_____	51–75
_____	76–100
_____	>100

33. Number of years you have been practicing:

27%	<5
27%	6–10
13%	11–15
0.2%	16–25
0.13%	>25

Appendix A

Endnotes

[1] Pragmatic validation requires that some other indicator be available that serves as an alternative to check the accuracy of sampling or observation. For this case, the evidence of actual constructed projects designed by professionals since the revisions allow a very real measure; do the constructed projects reflect the standard of care required in the code of ethics?

[2] Several questions were posed in different forms in the survey, allowing comparison of the answers. Similar questions should have similar answers and therefore support the validity of the question and answers.

Appendix B

The following is an excerpt from a personal correspondence with Xi Chunmei of Sichuan Fine Arts Institute in Chongqing, China, in which she outlines the relationship and reliance of professionals on published standards, etc. While professional groups have guiding principles, they tend to be focused on general obligations of the profession and less on individual obligations. The following excerpt is used with permission:

> Most of the institutions have their articles of association that mainly contain the qualifications and procedures of joining the institutions, the standard of membership fees, etc. Almost none of these nongovernment institutions has its own code of ethics; some of them just reprint relative laws or assessment standards from government institutions such as the Ministry of Housing and Urban-Rural Development of the People's Republic of China (MOHURD) http://www.mohurd.gov.cn/. For example, on the website of MOHURD, there are Environmental Protection Law of the People's Republic of China issued in 1989 and revised in 2014, Energy Conservation Law of the People's Republic of China issued in 1997 and revised in 2007, and Renewable Energy Law of People's Republic of China issued in 2005.
>
> MOHURD, together with the General Administration of Quality Supervision, Inspection and Quarantine of the People's Republic of China, has published some assessment standards such as Assessment Standard for Green Building issued in 2014, Assessment Standard for Green Retrofitting of Existing Building issued in 2015, and Design Standard for Energy Efficiency of Public Buildings issued in 2015. Besides, Architectural Society of China also published Assessment Standard for Healthy Building in 2017. Briefly speaking, these assessment standards mainly focus on the environmental aspect, paying little attention to other aspects such as recognition of human rights, protection of cultural heritage, nondiscrimination, and prohibition of unfair competition. The following table represents the contents of the assessment standards mentioned above.

Title	Time	Issued by	Contents
Assessment Standard for Healthy Building	2017	Architectural Society of China	10 chapters 1. General provisions 2. Terminology 3. Main provisions 4. Air 5. Water 6. Comfort 7. Fitness 8. Humanity 9. Service 10. Improvement and innovation
Assessment Standard for Green Building	2014	Ministry of Housing and Urban-Rural Development of the People's Republic of China (MOHURD), together with the General Administration of Quality Supervision, Inspection and Quarantine of the People's Republic of China	11 chapters 1. General provisions 2. Terminology 3. Main provisions 4. Land-saving and outdoor environment 5. Energy-saving and utilization 6. Water-saving and utilization 7. Material-saving and utilization 8. Indoor environment quality 9. Construction management 10. Operation management 11. Improvement and innovation

(Continued)

Appendix B

Title	Time	Issued by	Contents
Assessment Standard for Green Retrofitting of Existing Building	2015		11 chapters 1. General provisions 2. Terminology 3. Main provisions 4. Planning and buildings 5. Structure and materials 6. Heating, ventilating and air conditioning 7. Water supply and drainage 8. Electricity and gas 9. Construction management 10. Operation management 11. Improvement and innovation
Design Standard for Energy Efficiency of Public Buildings	2015		5 chapters 1. General provisions 2. Terminology 3. Calculation parameters of the energy-saving design of indoor environment 4. Buildings and the thermal design of buildings 5. Energy-saving design of heating, ventilation and air conditioning

Index

Note: Locators in bold refer to tables.

ACEC *see* American Consulting Engineers Council (ACEC)
act utilitarianism 52, 139
administration of examinations 22
affirmative duty 137
AIA *see* American Institute of Architects (AIA)
Alpern, Kenneth D. 35, 95, 109
Alpern's corollary of proportionate care 109
Alpern's Principle of Care 109
American Consulting Engineers Council (ACEC) 30, 33
American Institute of Architects (AIA) 30, 34
American Institute of Chemical Engineers 41
American Society of Civil Engineers (ASCE) 25, 33, 34, 36, 97, 148, 161
American Society of Heating, Refrigeration and Air Conditioning Engineers (ASHRAE) 14
American Society of Landscape Architects (ASLA) 25, 34, 148, 161
American Society of Mechanical Engineers (ASME) 41
American Standards for Testing and Materials (ASTM) 42
Anderson, Albert 1, 2, 5, 158
Anderson, Arthur 20
anthropocentric bias 58
anti-environment mission 3
anti-sustainability mission 3
Aristotle 19, 52, 158
artificial intelligence (AI) 125
ASCE *see* American Society of Civil Engineers (ASCE)
ASCE code of ethics 36–37
ASLA *see* American Society of Landscape Architects (ASLA)
assimilative capacity of environment 130
authority 21

balancing obligation and opportunity 73–76
Baxter, William F. 55, 56, 59
Bentham, Jeremy 52, 57–58
Berry, Thomas 7, 49
bisphenol-A (BPA) 127
Bok, Sissela 96
Boston Consulting Group 86
"boutique" firms 14
Bowie, Norman 97
Braungart, Michael 111, 147
Brundtland Commission 3
Building for Economic and Environmental Sustainability (BEES) model 153
business and sustainability 83–85
business ethics 49
business protocols 30

capitalism 60, 65
catastrophic failure 111, 112
categorical imperative 51
caveat emptor 129
Charter of the European Union 128
Clark, Colin 64
codes of ethics 22, 37, 76; due diligence 23–30; environment and sustainability considerations **38–40**; evaluation 30–34; peak organizations for design professions **26–29**; Sustainability and the Ethical Challenges 34–44
codes of professional ethics to due diligence test **31–32**
commercial real estate 64
Committee on Professional Conduct 37
community designer 114

175

community policy on the environment 128
community sanction 21, 22
compliance-oriented professional 14
consequentalist 49
consistency 134
constant capital rule 6
consumer consumption 151
consumption 6
Corollary of Proportionate Care 35
Corporate Eco Forum 3
corporate moral responsibility 88
corporate obligation 79–83
corporate stakeholders 82
corporations 79–80; act with intentionality 89
corporations aggregate 80
corporations sole 80
costs and benefits of nonaction 134
creative contributions 78
critical natural capital 7
criticism 120
culture 21

deductive reasoning 119
degree of care 35
degree of sustainability 68
degree of tolerance 30
Departments of Energy and Interior 3
design 131; defined 4; failure 112; professionals 2, 14, 15, 67, 92; values 151–157; *see also* design professionals
designing by standards 113
designing for sustainability 145
design professionals 145–149; as student 157–158; as teachers 149–151
design professional and organizations: balancing obligation and opportunity 73–76; business and sustainability 83–85; corporate obligation 79–83; deeds without doers 76–77; duty of corporation to professional employees 89–96; ethical agency 77–79; moral responsibility and corporations 88–89; sustainable business practices 85–88; whistle-blower 96–100

devaluation of things in nature 61
Diamond, Jared 6
dichlorodiphenyltrichloroethane (DDT) 132
dissemination of codes 34
doers 79
Donaldson, Thomas 74
due diligence test of corporate ethics programs 23–30
Durant, Will 73
duty of corporation to professional employees 89–96

ecological crises and sustainability 1
economic prosperity 3
economic wealth 5
Ehrenfeld, John R. 9, 111, 145
Elkington, John 85
employees 79
enforcement capacity 25
Engineers Council for Professional Development (ECPD) 22, 93–94
environmental and sustainability ethics 85
environmental concerns and sustainability 85
environmental functions 151
environmental protection 20
Environmental Protection Agency (EPA) 3, 153
environmental values 150
Epimethean philosophy 2
Epimetheus 1
equity 3
ethical agency 77–79
ethical codes 21
ethical judgments 68
ethical obligations 49, 84, 107, 110; of design professionals 43; nature as property 63–64; other living things 54–55; private property and sustainability 65–67; professional standards 68; society's worldview 60–63; speciesism 57–60; theories of ethics 49–54; utilitarian views of nature 55–57; value of property and nature 64–65

Index

falsifiability 121
Firmage, Allen D. 22
Florman, Samuel C. 35, 43, 108
Foundation of Professional Practice 34
framework of regulations 107
fraud (falsification of education) 30
Freeman's Doctrine of Fair Contracts **75**
free-market ideal 21
"free rider" problem 23
Friedman, Milton 81, 82
futurity problem 138

General Services Administration (GSA) 154
genetically modified (GM) food 130
Global Reporting Initiative (GRI) 87
global warming 130
Gokany, Indur M. 132, 133
Goodpaster, Kenneth E. 76, 90
Gould, Stephen Jay 49
Green buildings 14, 151, 152
"greener" projects 15
Green Revolution 6
Greenwood, Ernest 94
gross domestic product (GDP) 64
Guidelines to Practice Under the Fundamental Canons 37

Harding, Garret 77
Harrison, George 145
health and the environment 3
Hermes 1
hippocratic oath 67
homo economicus 5
Housing and Urban Development (HUD) 153
housing markets 12
housing real estate 64
human rights 54
hypotheses 121

incentivize conservation 5
individual design professional 92
inductive logic 120
inductive reasoning 120
Industrial Designers Society of America (IDSA) 41
information dissemination 25

Institute of Electrical and Electronic Engineers (IEEE) 41
Institute of Engineering Ethics Education (IEEE) for engineers 99–100
institutionalized class systems 50
intrinsic natural rights 130

Jackall, Robert 78
Jackall's organization 79
Jardins, Des 57
juristic person 80

Kant, Emmanuel 51, 52, 55, 83
Kantian Categorical Imperative 51
Kew Gardens guidelines 117
Kew Gardens principles 73, 74, 95, 109, 136
key performance indicators (KPIs) 87
Kuhn, Thomas S. 119, 122, 146

Ladd, John 34
laissez-fair 81
laissez-faire system 80
land development 11; in the United States 64
Langdon, Davis 154
Leadership in Energy and Environmental Design (LEED) program 12, 111, 153
Leopold, Aldo 127
licensure 19, 20, 21; criticism of 20
lifestyles 12
limited liability 80
Locke, John 55, 59, 60, 61, 63
Lovins, Amory 7, 146

Maastricht Treaty 128
Maitland, Ian 23
Malthus, Thomas 6
market 19
Maslow, Abraham 4
Maslow's Hierarchy of Needs 50
Massachusetts Institute of Technology Business Review 86
McDonough, William 7, 8, 9, 15, 107, 111
McFarland, Michael C. 73, 74, 95, 109
Medical Ethics 90

Mill, John Stuart 53, 139
Mill's Greatest Happiness Rule 53
Mill's Harm Principle 139
moral agent 79
moral behavior 147
moral character of organization 78
moral choices 79
morality 51, 53
moral minimum of corporations 82
moral responsibility and corporations 88–89
moral rights 54
moral rules 50
moral standards 54
Morris 130
Morris, Julian 131
multinational corporations 83

Naruto 59
National Institute of Standards and Technology (NIST) 153
National Society of Professional Engineers (NSPE) 41
natural persons 80
natural resources 5
The Natural Step 87–88
needs 74
negative externality 63
negative human rights 54
Nemo Dat Principles (NDP) 76, 90
nonconsequentalist 49–50
nondiscrimination 134
nonlinear system 85
normative theories of ethics 49–50

Obama, Barack 3
obligations to care 130
occupational licenses 19–20
organizations convicted of federal crime 24

peak organization 23
performance-based system 114
personal liability 81
personal service 20
Petroski, Henry 111
plagiarism 25
Plato 1
Popper, Karl 119, 121

practicing designers 8
precautionary obligation 141
precautionary principle (PP) and design 127–129, 135–137; duty to prevent harm 137–140; for future 137–138; precaution and design 140–141; underpinnings 129–135
President's Council on Sustainable Development (PCSD) 3
preventative anticipation 129
prima facie obligation 50
Principle of Care 35
private organizations 21
profession 19–23
professional authority 21, 22
professional conduct 33
professional development 30
professional employees 89–96
professional ethics 49, 50
professionalism: attributes 22; defined 22, 34; element of 93; moral responsibility 95
professional licenses 19–20
professional obligations 127, 137
professional organizations 25
professional peers 33
professional practice standards 68
professional self-regulation 23
professional's obligation 24
Promethean ethos 145
Promethean philosophy 2
Prometheus 1
property 62
proportionality 130, 134
Protagoras 1
proximity and capability 74
public policy 35, 64
public safety 20
public sanction 19
public transportation 12

quality 110–111
quality-of-life issues 109

Rawls, John 51, 140
Rawlsian argument 75
reasonableness 137

Index

"reasonableness" test 146
resource management 5, 6
reviewers 115
right of self-determination 97
risk assessment 131
rule utilitarianism 139

Sagoff, Mark 65
scientific method 121
scientific revolutions 122
self-interested individuals 52
self-regulation 23
self-serving behavior 23
short-term environmental impact 63
Simon, John G. 73
Singer, Peter 57
Slater, David 59
Smith, Adam 61, 80, 81
social equity 3
society sanctions 22
Society's Code of Ethics 37
Spier, Raymond 43
stakeholder-based approach 76
stakeholder shortlist 50
standard-of-care concept 146
standards 15
state-sanctioned license 20
Steinberg, Theodore 63
strong-side sustainability 6, 7
suburban development 10
supervisors 79
sustainability 62, 86, 107–108; and design ethics 122–125; design professionals 108–110; flourishing 145, 156; and obligation 116–118; practices 14; science and design 118–122; standards and 110–116; weak 66
sustainability and design 2; defined 3–9; need of 9–12; paradigms of 3; selling 12–16; unsustainable 2–3
sustainable business practices 85–88
sustainable design 15, 108, 152
sustainable development 36, 41; goals of 2–3
sustainable society 4, 87
systematic theory 21

Ticknor, Joel 132

Unger, Stephen 90
United Nations Conference on Sustainable Development 128
universalizability 83
utilitarian ethics 52
utilitarianism 52, 53, 139
utility 51

value 62
virtue ethics 52
virtue morality 51

weak-side sustainability 6, 7
Wealth of Nations (Adam Smith) 88
well-being 107
whistle-blower 96–100
Wilson, Edward O. 65, 66
Wingspread Conference (1998) 128

zero-sum approach 4
zero-sum game 56
Zeus 1–2